What Makes You So Special?

From The Big Bang to You

Rosa Rubicondior

What Makes You So Special?

Cover photo: Vicki Ward.

Keira Mai and Mia Rose Hounslow.
Two very special little people.

Taken at "Colourscape", Waddesdon Manor, Waddesdon, Buckinghamshire.
June 2017

June 2021 Minor revisions and re-indexing

April 2025 Minor revisions.

ISBN-13: 978-1546788294

What Makes You So Special?

When I became convinced that the Universe is natural-that all the ghosts and gods are myths, there entered into my brain, into my soul, into every drop of my blood, the sense, the feeling, the joy of freedom. The walls of my prison crumbled and fell, the dungeon was flooded with light, and all the bolts, and bars, and manacles became dust. I was no longer a servant, a serf, or a slave.

There was for me no master in all the wide world-not even in infinite space. I was free: free to think, to express my thoughts; free to live to my own ideal; free to live for myself and those I loved; free to use all my faculties, all my senses; free to spread imaginations wings; free to investigate, to guess and dream and hope; free to judge and determine for myself; free to reject all ignorant and cruel creeds, all the "inspired" books that savages have produced, and all the barbarous legends of the past; free from popes and priests; free from all the "called" and "set apart"; free from sanctified mistakes and holy lies; free from the fear of eternal pain; free from the winged monsters of the night; free from devils, ghosts, and gods.

For the first time I was free. There were no prohibited places in all the realms of thought, no air no space, where fancy could not spread her painted wings; no chains for my limbs; no lashes for my back; no fires for my flesh; no masters frown or threat; no following another's steps; no need to bow, or cringe, or crawl, or utter lying words. I was free. I stood erect and fearlessly, joyously, faced all worlds.

Robert G. Ingersoll (1833-1899)

What Makes You So Special?

Dedication

This book is dedicated to our ancestors, each and every one of them, without whom none of this would be possible.

They never failed us. We owe everything to them.

It is also dedicated to my late sister–in–law and dear friend, Therese Al Sharhan, who sadly died when this book was nearing completion.

What Makes You So Special?

Acknowledgements

I would like to thank all those on the social media who encouraged me to expand on the brief outline of science, originally published in 2010 on my Rosa Rubicondior blog, and to work it up into an up-to-date book.

My thanks also to my long–suffering partner who again had to put up with me engrossed in writing, reading and grunting out monosyllabic, often inappropriate, answers in place of conversation.

I would also like once again to thank my sisters in law for the use of their cottage in the country when I needed solitude to concentrate.

What Makes You So Special?

Contents

What Makes You So Special?

Introduction

This is your story. It tells how you came to be here and why you are unique.

Bear with me because yours is a long, long story even when compressed into less than 200 pages, but it is a story very much worth telling.

The really great thing about your story is that, to borrow from Carl Sagan, you share it not only with everyone you know; everyone you ever heard of; everyone who ever lived on this planet, but each and every living thing that ever lived here in this tiny dot in the cosmos.

This book is based on a blog post I wrote several years ago (Rubicondior, Rosa, 2010). Several people have suggested I work it up into a book aimed at an interested but not especially academic readership. That post was itself based on a posting in the CompuServe SciMath forum which I wrote in a single evening, and which was met with general acclaim.

There have been many advances in science, especially in our understanding of human evolution since I wrote the original some twenty years ago. For example, had I drafted this book then I would have said that there is only very tentative and not universally accepted evidence that modern humans ever interbred with Neanderthals. Now we know they not only did but that they also interbred with other archaic hominids, some of which have not yet been identified.

The purpose of the book is to tell the story of the Cosmos, a kind of history of everything, from the point of view of the reader and explaining how and where we fit in the general scheme of things. I have tried to avoid too much technical detail, confining more detail to endnotes where necessary, to keep the narrative flowing along without too many diversions.

It is not a comprehensive review of science nor of the evidence behind generally accepted scientific theories, but hopefully I have included enough detail to stimulate interest and to encourage a deeper understanding. The study of science does not require a university degree to make it enjoyable. Science is an adventure waiting to be had. Some of it is not really hard to understand so much as hard to believe.

I am not happy to just look at a rose and marvel at its beauty of form and fitness for purpose; I want to know why it looks like it does and smells the way it does. I want to know how it makes the colours and scents, and I want to know why this is attractive, not just to the bees and other pollinating insects, but why it is attractive to us. I do not want to inhabit a Universe where there are mysteries we must accept as forever closed to us and beyond our understanding, as though having a mystery is an answer and believing in mysteries and magic is regarded as wisdom. I want to know. I want to experience Richard Feynman's joy of finding things out.

I hope this modest little contribution will stimulate that desire in others and will raise more questions than it answers. I want to stimulate exploration and discovery.

As always with my books, please never take my word for anything, but check for yourself. Expecting you to agree with me would be pedagogy of the most arrogant kind. Question everything and never settle for dogma. The only certainty in science is that there are no certainties.

Above all, enjoy your story. Yours is a life worth living.

1. Something for Nothing

The best place to start with the best of all stories is the very beginning and the first amazing fact is that your story started at the very beginning of everything. As we'll see, I can't even start it with "Once upon a time", because it starts at the very start of time itself. Your story started when the Universe started, about 13.8 billion years ago when the Universe was something astrophysicists like Stephen Hawking call a singularity – although I'll have more to say about that later.

There are two important pieces of evidence that the Universe started out very small and came into existence in a single event about 13.8 billion years ago. This is colloquially referred to as the Big Bang – possibly the worst misnomer in all science because it was neither big nor went bang. It was a small silent event; silent because with no space and no matter there can be no sound. It was actually coined by the astronomer, Fred Hoyle to mock the idea. Fred Hoyle was an advocate of the 'steady state' theory which argued that the Universe had always existed and went through cycles of expansion and contraction.

The first piece of evidence was the discovery by Edwin Hubble that the light coming from distant objects is shifted towards the red end of the spectrum, showing that the Universe is expanding. The Red Shift effect is rather like the effect you get as a police car siren changes tone as it goes past you or the noise of a non–stop train changes as it goes through the station. It is higher pitched when coming towards you, and lower pitched when going away. The sound waves are compressed as the vehicle comes towards you and stretched out as it moves away, because the source is moving. The sound is 'shifted' depending on whether the source is moving towards you or moving away. This effect is known to science as the Doppler Effect.

The red end of the visible colour spectrum has a longer wave–length than the blue end, so the Red Shift tells us the objects that the light is coming from are moving away. The further away objects are the more they are red–shifted. This means objects further away appear to be moving faster than near objects.

Knowing how far on average light is shifted toward the red end of the spectrum means we also know how fast these objects are moving away from us. From that, cosmologists can calculate that they all started out in a single point, about 13.8 billion years ago.

In actual fact, this is not quite the true picture because the Red Shift can be shown to be the same for any point in the Universe. If everything were radiating out from a centre this would only be true at this central point. So, from that we know that it is not that things are flying apart but that the space between them is increasing. The Universe is expanding because space is still being created. The difference is small but significant.

To understand this, think of dots on the surface on an inflating rubber balloon. The more the balloon inflates the further away from each other the dots are. However, it is not the dots that are moving but the space between them that is increasing as the rubber stretches. They are not radiating out from a central point; rather, the rubber of the balloon is stretching between them.

Now, translate that into three dimensions and imagine the dots as embedded in a matrix, like suns and galaxies and other celestial bodies embedded in space. Whichever body you pick as your point from which to observe all the other bodies, the further away the bodies the faster they will appear to be moving away at a rate directly proportional to their distance away from you. A body twice the distance away as a close body will have twice as much space expanding between it and you as the nearer body has.

And this is true for every point in the Universe!

This means that the Universe does not have a centre! The 'centre' – the point at which it all started – is everywhere. It is this that is expanding. The cause of the Red Shift is the same, however. As the space between the objects increases so the light waves are stretched and shifted towards the red end of the spectrum.

The second piece of evidence that everything started in a huge generation of energy is that we can still detect the 'echo' of this. At about one ten thousandth of a second after the Big Bang the Universe would have been extremely hot – about one trillion degrees Celsius. This heat was high–energy electromagnetic radiation. As the Universe expanded this would have cooled to about ten billion degrees after one second and about one billion degrees after three seconds. It would continue to cool but would never reach zero.

This remnant electromagnetic radiation is now the cosmic microwave background (CMB) and it is exactly what we would expect it to be given the age of the Universe calculated from the Red Shift. This is the 'smoking gun' of the Big Bang. The CMB was discovered by accident by American radio astronomers, Arno Penzias and Robert Wilson who received a Nobel Prize for their work (Evans, 2015). They initially thought it was a fault with their equipment, even scrubbing their radio antennae to clean pigeon droppings off them in case that was the problem.

Having eliminated everything they had to conclude that there really was something causing a hiss in the microwave wavelength, and it was everywhere they looked.

Ironically, the CMB had been predicted by Fred Hoyle, who taunted Stephen Hawking, with whom he disagreed about the Big Bang, by cited the absence of it as evidence against the Big Bang.

Although he didn't realise it at the time, Albert Einstein's Special Theory of Relativity predicts that the Universe should be expanding. He had realised that his first version of his formula showed this but had assumed there must be another factor, which he called the 'cosmological constant', to correct for this 'error'. He later called this the greatest mistake of his life. He had made the fundamental mistake of allowing his intuition to over–ride the evidence! Had he had the confidence to stick with his original formula he would have predicted the Big Bang.

In Einstein's defence, it was not generally understood in the early 20th Century that the Universe was as big as we now know it to be. It was thought that our galaxy – the Milky Way galaxy – was the Universe. Astronomers did not know then that the Milky Way is but one of maybe half a trillion galaxies. A Universe as small as Einstein thought it was, and expanding at the rate his formula predicted, could not be nearly old enough.

It was for another mathematician, the French Jesuit priest, Georges Lemaître, to work out from Einstein's formula that the Universe was expanding so it was he who first predicted the Big Bang. So excited was the then Pope, Pope Pius XII, by this discovery that he declared this was the "Let there be light!" moment in the Bible and declared the Bible to have been scientifically proven to be true. Lemaître was scientist enough to realise that the Pope had tied 'proof of God' to a scientific theory and that scientific theories could be falsified and often were by subsequent discoveries. Had Einstein's Theory of Relativity – the very basis of the Pope's proclaimed proof of God – not falsified Newton's Law of Universal Gravity? After his advice, the Pope went quiet on the subject and the Vatican has never explicitly repeated that claim, although it does not refute the Big Bang.

So, there is pretty much universal acceptance in science circles that the Universe started out as a sudden, spontaneous event that we call the Big Bang. Stephen Hawking and the mathematician Roger Penrose together showed that this would have been a 'singularity' such as there is at the centre of a Black Hole. As I will expand on later, there is now some dispute about this because a singularity is a prediction of Einstein's General Theory of Relativity[1] where mass is so great that it distorts spacetime to give a gravity well from which nothing can escaped, not even light. But, a singularity is, by definition, very small and so comes under the Quantum Domain, not the Relativity Domain For the moment though I will talk about the singularity.

A singularity is not **just** something exceedingly small, although it **is** exceedingly small. It is something so small that it is as small as it is possible for something to be without quite being nothing. A singularity is so small that there can be no fundamental laws governing it. The fundamental laws describe how things react together, transfer energy between one another and generally get along. But if there is only one thing, there is nothing to describe, and so no laws.

Laws of nature, of course, are descriptions of what happens, not some sort of instructions that have to be followed. They are what we call descriptive laws, not proscriptive or prescriptive laws. This an important distinction which often confuses people into believing laws of nature mean there must be a law–maker and even some sort of enforcement.

Imagine if you will, cars being driven along Highway 15, north of San Bernardino, California, when the Pacific Plate, west of the San Andres Fault slips a few more yards north relative to the North American Plate on the east side of the fault, and the road surface tilts 90 degrees. Suddenly, the cars, the drivers of which had been dutifully obeying the proscriptive and prescriptive motoring laws of California, suddenly start obeying the descriptive natural Laws of Motion instead.

There won't be any traffic police to ensure they continue in a straight line unless acted upon by a force, or to ensure that when two cars travelling in opposite directions collide they come to a sudden stop but the occupants try to continue travelling in the same direction until they meet an immovable object. No–one needs to tell them what to do; they just get on with it and do what the descriptive laws describe and predict.

Now back to the singularity. The singularity cannot have had a cause because there was nothing to cause it. It could not have existed anywhere because there was nowhere for it to exist. It cannot have had a before because there was no time for it to have been in. It did not exist in space or time because there was no space or time for it to exist in.[2]

For all practical purposes, it did not exist. It is zero; nothing; nada; not.

And yet this nothing gave birth to the entire Universe!

To begin to understand this you need to begin to understand quantum physics. (Polkinhorne, 2002) (Don't panic! There won't be a test, and I don't understand it that much either!).

There are a few basic things to understand about quantum physics:

1. Quantum physics is very strange and counter–intuitive. Things do not behave the way we expect them to in Quantum World and the laws we have got used to in Big World do not apply there. Quantum World is a world where we have to accept that our human intuition is a very poor tool for measuring reality. It is a very poor tool for that anyway, as Albert Einstein discovered to his cost, but even more so in Quantum World. It is so strange that if you think you understand

quantum mechanics, you very probably do not understand quantum mechanics.

2. There are no absolute values in Quantum World. Quantum values fluctuate and wobble around an average value. This is because Big World values are an emergent property of Quantum World chaos. This also applies to zero. There is no such thing as nothing in Quantum World. A value can fluctuate around zero, sometimes less than zero and sometimes more. The range of this fluctuation in unbounded. It is more like a bell–curve than a straight vertical line on a graph. So, although it always averages zero, 'nothing' can be plus lots or minus lots.

3. If a particle **can** exist in more than one state it **will** exist in all possible states simultaneously. Scientists refer to this as a superposition of histories.

4. Quantum theory leaves you scratching you head and thinking that cannot be right, but when they do the maths and the experiments, scientists find that it **is** right. What was wrong was our intuition. Quantum World is weird (see point 1 above).

But, out of this strange, counter–intuitive, tiny little fluctuating nothing, the events that were to result in you had their beginnings.

It was at that instant that space and time began to exist. At that point there was somewhere and somewhen for you to eventually exist in. Space and time are part of the Universe. Outside the Universe there is no space or time and so nowhere for an 'outside' to exist in. Literally everything that is part of this Universe is in this Universe. There may be other Universes, but they exist in their own space and time.

Astrophysicists believe the smallest quantity of time that can exist is the so–called Planck time. This is 10^{-43} seconds or 0.0000000000000000000000000000000000000000001 seconds. This means that immediately the universe came into existence it was already 10^{-43} seconds old. This time, for our universe, was enough for gravity to be stripped off from the other three forms of energy

– weak and strong nuclear forces and electromagnetic force – which are the forms of energy of which, in one manifestation or another our Universe is composed.

So, a tiny fluctuation around zero time could have given us a Universe that was already old enough to enter what astrophysicists call hyperinflation during which there was a massive increase in size from zero as space was created.

Now, you might be wondering how there can have been a time with nothing before it. How can time itself have had a beginning when all our intuition tells us there must have been something earlier?

To understand this, imagine yourself trying to find where south starts. Intuition says you need to walk away from south because when facing south, south is in front of you; so south must have started behind you. So, you turn round (or walk backwards) and walk north as far away from south as it is possible to walk – to a point where you cannot walk any further north. You would find yourself at the North Pole where every direction is now south! There would be no more north to walk towards and no more south to walks away from.

There is nothing north of the North Pole and all of south leads away from it. This is S_0 – zero south. It is the same with time; there is nothing earlier than the Big Bang because the Big Bang was where time started just as the North Pole is where south starts. Asking what was before the Big Bang – before T_0 – is as meaningless as asking what is north of the North Pole.

The fact that this is a hard concept to grasp, unlike the idea of there being nothing north of the North Pole, is because our intuition is not the best measure of reality. We have evolved in an environment in which there was always something earlier and our mind assumes there always must have been. The maths, however, tells us this is not necessarily so and could not be so at T_0.

Some things in science, like Relativity and Quantum Theory, are not really that hard to understand; they are just hard to believe. Most people have no difficulty with the idea that there is nothing north of the North Pole, although I have had one indignant American fundamentalist tell me that Russia is north of the North Pole because if you walk north from the USA you eventually get to

Russia. Apparently, it was my ignorant stupidity that meant I couldn't understand this! And she was serious!

So, cosmologists refer to this time at the instant of the Big Bang as T_0 or zero time, but it is even more fundamental than that. If there was no time there was no space either because, as Albert Einstein worked out, space and time are part of a continuum – spacetime. With no time there is no space. But then for some, the idea of a spheroid Earth is a difficult concept too. It looks more or less flat, so it must be, give or take the odd mountain or valley.

But how can this possibly be? You see, our human intuition tells us this cannot be true; that there **must** have been something for the Big Bang to happen in. There must have been something, even nothing, before there was something. But of course, we aren't used to thinking of things without time. We take time so much for granted that we assume it has always been there; that there was never no time. It is not hard to understand; it is simply hard to believe.

We are also used to thinking in terms of nothing, or what we imagine nothing to be. We have nothing in the fridge until we have done the shopping. There is nothing on a new, clean sheet of paper until we draw on it. There is nothing worth watching on TV tonight.

But none of these were really 'nothing'. The fridge didn't have 'nothing' in it; it had empty space (or more precisely, air). The sheet of paper didn't have nothing on it; it had a surface which was made of paper; and the TV was still a TV.

In fact, when you think about it, how can there ever be nothing and why do we assume there ever **was** nothing? What exactly **is** nothing anyway? Can we describe it or measure it, or discover anything about it? Can 'nothing' really exist? And why assume the default state of existence is non–existence? What sense does that make?

Surely, by asking how the Big Bang came from nothing we are assuming there ever was nothing in the first place. But isn't that a logical absurdity? If nothing is non–existence, how can nothing have existed? How can 'nothing' both exist and be non–existence at the same time?

In fact, of course, we are tangling the idea of nothing with the idea of 'before' and yet there was no time before T_0. So why are we assuming this 'nothing' before something anyway?

Can 'nothing' really exist? Of course not, and nor can we say anything about 'nothing' because there is nothing to say about it. When creationists and other theists argue that nothing can come from nothing, they are claiming to know something they cannot possibly know. They are claiming to know about 'nothing', and yet they can know nothing about it. They are making an intuitive guess from ignorance, as though an intuitive guess alone wasn't bad enough.

So, logic tells us that the singularity was not something from nothing but something from something. There are several ideas about what could give rise to a Big Bang, and so to a Universe. One of them is a Black Hole where matter is so condensed under its own gravity that it is actually a singularity. Some astrophysicists think the conditions inside a Black Hole are right for the birth of Universes. Stephen Hawking explained how this could happen in his book, *Black Holes and Baby Universes* (Hawking, 1994).

Some astrophysicists believe the nearest thing to 'nothing' that can exist is a zero energy quantum vacuum.

And some astrophysicists, like Stephen Hawking himself, now think the Big Bang at T_0 might not have been a singularity in the classic sense anyway because a singularity is a prediction of Einstein's General Theory of Relativity. In other words, it is a prediction of Big World. But, below a certain size (and scientists aren't sure exactly what that size is yet) we enter Quantum World and the Big Bang at T_0 was firmly in Quantum World.

So, it looks increasingly likely that the Big Bang was a fluctuation in a non–zero quantum energy field inside a Black Hole. This view might change as more is discovered but it doesn't change the facts of what happened next.

Recall what I said about quantum zero earlier? Quantum zero is not an absolute value but only an average value for an unbounded range of fluctuations, both positive and negative, so, rather than a quantum vacuum being zero energy (which it is on average) it is actually a fluctuating non–zero quantum energy field.

And non–zero quantum energy fields **can** give rise to spontaneous particles arising, apparently, out of nothing because, as Einstein showed with his famous $e = mc^2$ formula, energy and mass are interconvertible.

Then we have stuff astrophysicists call 'quantum foam'. This idea comes from the observed phenomenon that at the quantum level, all possible states for a particle exist simultaneously and that what we see is actually a 'superposition' of all these possible states. Quantum physicists call these different states 'histories'. A simple, and quite spooky example of this is the 'two slit' experiment, where a single photon fired at a metal plate with two slits in it can be shown to pass through **both slits** at the same time and so make an interference pattern with itself on a photographic plate behind the two slits!

When you realise that a photon is a wave as well as a particle, this makes a little more sense, but the two slit experiment gets really weird when scientists set up a detector over one slit to 'watch' the photon pass through. It then doesn't pass through the other slit! The act of observing it causes the wave form to collapse, and it becomes a perfectly normal particle!

In fact, physicists now talk in terms of a particle 'field' rather than a wave–particle duality. We know that particles are also waves, but that doesn't seem to make sense, does it! And why does the wave collapse when we observe it? If you think you understand quantum mechanics, you almost certainly do not understand quantum mechanics. It is that intuition thing again! The observable evidence tells us one thing; our intuition tells us another. The evidence wins.

So, in our non–zero quantum energy field, with our singularity or whatever the Quantum World version is, free to fluctuate around zero and having all possible histories, there was nothing to stop it forming an infinite number of universes with all possible conditions! This was the quantum foam of baby universes. There are two ways of looking at this. You can either see these as lots of different Universes, or one Universe with lots of different 'histories'. It doesn't really matter very much to your story because we are dealing with this Universe and this Universe's history.

Some of these universes would have had conditions which caused them to collapse immediately; some might have collapsed after a few hours, days, months or years and some, like ours could have had (in fact we know one **must** have had because we live in it) the right conditions to carry on expanding, to form particles, atoms, suns, galaxies and planets, at least one of which was right for the eventual evolution of life.

So, as soon as the Universe began to exist, it was already 10^{-43} seconds old and the conditions were right for lots of negative energy and lots of positive energy to be created, but always adding up to zero energy. Nothing was literally becoming something and your story had begun.

And we still do not know exactly what happened during that first 10^{-43} seconds. Maybe we never will; maybe the best we can ever do it to produce plausible mathematical models for it. For a technical but very readable book explaining how a Universe can apparently come from nothing, see *A Universe From Nothing: Why There Is Something Rather Than Nothing* by Lawrence M. Krauss. (Krauss, 2012)

Some people find it hard to believe that the conditions in this universe were just right for us to evolve in. They do not understand how this need not have been intended because they find it hard to comprehend a Universe without them in it and try to calculate the probability of it happening. They then conclude that it was too unlikely to be by chance alone.

This is because they do the wrong sums, as I will show in the next chapter.

This anthropocentric view of the Universe – the notion that it is all for us – is based on nothing more than the religious claim that a god created it as somewhere for humans to live. There is no scientific evidence for that notion at all, just as there is no scientific evidence that there is some ultimate purpose; that the Universe is **for** something.

What Makes You So Special?

End notes:————————————————

[1] It is useful to think of physics as dealing with two domains – Relativity, which deals with the very large; and Quantum which deals with the very small. So far, scientists have not been able to devise a 'Grand Unifying Theory' which integrates the two. This is considered by some as the Holy Grail of physics.

[2] The Universe still doesn't exist **in** anything. 'Outside' the Universe there is no space and no time because they only exist 'inside' the Universe, so there is nowhere and no time for an outside to exist in. The Universe has expanded only into itself. This has another counter–intuitive consequence – there is no edge to the Universe; it doesn't have a boundary. The Universe is unbounded and yet it isn't infinite. To describe something as existing outside of space and time is just another way of saying it doesn't exist

.

2. Why You?

In Chapter 1 I covered just the first 10^{-43} seconds and how something came from nothing. By one ten thousandth of a second there was so much stuff in our Universe that it was one hundred trillion times as dense as water and had a temperature of about one trillion degrees Centigrade and our Universe was in a run–away inflation as spacetime was being created.

We now need to look at what happened next, but before we do, I'll explain why I said in the previous chapter that creationists get their sums wrong when they try to calculate the probability of the Universe being as it is, and so of us existing in it.

Firstly, they cannot know any of this because to calculate these probabilities they need to know the probability of all possible other conditions in all possible other universes and then show that this Universe **could** have been different. As a simple analogy, to be able to calculate the probability of dealing any particular single card in a pack of cards you need to know how many cards there are in the pack. Creationists are in the position of not knowing if there is one or an infinite number of cards! All they have is a guess and, no doubt by an amazing coincidence, their guess just happens to be exactly what they need to 'prove' their point.

So, until they can show that, say, the velocity of light in a vacuum could have been different they cannot show that it could have been different in this Universe. For all they know there could be no possibility of it being anything other than 186,000 miles per second. In that case it wasn't highly unlikely at all, but certain. It is analogous to dealing a card from a 'pack' containing only that card! The probability of dealing that card is 1. In other words, it is certainty. It is not in the least unlikely. In fact, it would defy the laws of mathematics if it were anything else.

The same goes for all the other 'parameters'. Creationists do not know what the probability of them having just the value they actually have is. They do not even know if there is a possible range of values. They simply assume the probability must be very small then they multiply them all together to get an infinitesimally small value. This is the scary number tactic. You are supposed to think it is next to impossible.

There is also a theologically embarrassing assumption behind the claim that the Universe had to be just right for human life to exist in it. The assumption is that a putative creator could only work within this very narrow range of parameters. Yet isn't a supreme creator supposed to have created these conditions and so to be unconstrained by them? If it can only create life in very precise conditions, it is not omnipotent and is constrained by some other force or limitation. Who or what set up those constraints and imposed this seemingly immutable limitation on this putative creator? Like so many other religious apologetics, this one spirals into an absurd regress of infinitely greater gods and finally disappears up its own absurdity.

The 'finely–tuned' creationist argument is theologically contradictory, and mathematically false, as I will show in a moment, but that is not a problem for us. Creationists and theologians have enough to worry about without having to cope with a creator who they need to be supremely omnipotent and constrained by some higher power at the same time for their 'arguments' to work. We are talking here about what can be measured, observed or logically deduced from the evidence.

So why do they get the maths wrong when calculating these probabilities? The problem is they are trying to calculate the probability of a past event whose outcome is already known.

The late Nobel Laurate physicist, Richard Feynman, used to illustrate this point to his students by starting a lecture by saying "I can hardly believe it! There is a car in the carpark right now with the registration number [some registration number or other]! Isn't that amazing! I wonder what the probability of that is!" The student would then mentally try to calculate the probability of that car being in that carpark, right at that moment, with exactly those letters and numbers…

Intuitively, the probability seems vastly small – so small as to be almost impossible – but it is actually 1^1 – certainty – because Feynman simply picked one of many registration numbers on a random car in the car park. There were always cars in the carpark!

Allow me to illustrate this more with a couple of mind experiments:

Firstly, and you can do this for real, or as a mind experiment if you wish, take a pack of cards, shuffle it well and deal four bridge hands of thirteen cards each. Notice how easy that was? It didn't require any exceptional skills or magical powers. Unless disabled, anyone can do it.

Now calculate the probability of dealing exactly those hands.

It actually doesn't matter what the hands were, the maths is the same. The number of possible deals is mathematically expressed as $52!/(13!)^4$. For the non–mathematical, the expression $52!$ (factorial 52) means 52 x 51 x 50 x 49.... etc. ... down to ... 3 x 2 x 1. Factorials very quickly get very large. This formula means the probability of you dealing exactly those four hands was 1 chance in 53,644,737,765,488,792,839,237,440,000.

Do the same again. To calculate the probability of two events like this we multiply the two individual probabilities together, so the probability of you having dealt that first four hands followed by that second four hands is now 1 in $(52!/(13!)^4)^2$. This is 1 in 2877757889928060000000000000000000000000000 000000000000000000.00. Is that vanishingly small enough yet? If not, deal again two more times and calculate again. This time you need to multiply the answer above by itself. In a few more deals this number will be so large that it would fill this book and probably more besides just to write it out fully. A few more and it will be larger than the number of particles, including neutrinos, in the visible Universe, reckoned to be about 10^{86}. How many deals will it take for it to become indistinguishable from zero, meaning it was impossible for you to have dealt those hands?

And yet you have just dealt those hands without the assistance of magic and by doing nothing special other than the simple task of shuffling and dealing. No unique skills or powers required. You see how easy it is to produce these big scary numbers – and how fallacious is the reasoning?

Creationists who argue that the possibility of this universe existing is so small that it must have been done by magic are arguing, in effect, that it is impossible to deal a few rounds of bridge without magical powers which no human or natural process could possess.

They are of course trying to calculate the probability of a particular event – one that has happened. The probability of an event happening that has already happened is not small; it is 1, i.e. certainty. No bookmaker is going to give you odds on the winning horse winning last year's Kentucky Derby or Grand National because we already know the result and no gambler in his right mind would bet on a different horse winning. Quite simply, calculating the probability of a past event when the outcome is known is meaningless.

Consider the situation right now. You are reading this as an intelligent human being on a planet orbiting a sun in a universe in which you **must** exist and in which this planet and this sun **must** exist, otherwise you could not be reading it. It must also be a universe in which I exist for you to be reading what I wrote.

Now, you may not agree with me about religion but you cannot disagree with those basic facts. We both **must** exist in a universe in which intelligent human life exists and in which this intelligent human life is able to communicate via the written word, otherwise you could not be reading what I have written. The fact that you are reading what I have written means we both must exist in a universe in which we **can** exist.

A universe in which intelligent human beings can discuss the probability of such a universe existing **must** be one in which intelligent human beings can exist. This is the 'anthropic principle' and, as far as I can see, the logic is inescapable.

It matters not what the possible conditions in other possible universes are or were. It matters not what, 13.8 billion years ago, was the probability that you would, 13.8 billion years later, be reading something I wrote. The fact is that this is the hand that has been dealt. Neither you nor I were the intended outcome, but we are here and now. Just as those hands of cards were not the intended outcome, so the actual hands have no bearing on your ability to deal them. And our existence has no bearing on this Universe's ability to produce just us. It produces human beings, and billions of them!

Well yes, but why **you** in particular?

Try this mind experiment.

Imagine you could print the names, in no particular order, of every living person, a hundred to a page in books with a thousand pages each. That is one hundred thousand people to a book. The current estimate of the human population of Earth is 7.5 billion, so that is 75,000 books.

Line those 75,000 books up in bookcases on two opposite walls with half on the right and half on the left.

Now toss a coin. If heads pick the right–hand bookcases; if tails, the left–hand bookcases. Toss again and pick the left-hand half of the bookcases. Continue until only one bookcase is left.

Toss the coin again. If heads pick the top half of the bookcase; if tails the bottom half. Keep tossing and selecting half the books at each toss and discarding the other half, until you have one book.

Toss the coin and choose the first half of the book if heads; the second half if tails.

You're probably getting the idea now, so continue tossing coins to select the final page, the final half page, and so on, until you have just one name left.

What was special about the person? Nothing at all, of course, and yet every toss of the coin had to have the right outcome for him or her to be selected. It needed a run of 23 correct coin tosses to pick just that one person! The probability of that? 1 in 7.5 billion of course – another big scary number. Surely there must be something special about a person for whom 23 consecutive coin tosses all have to be right, isn't there?

Well, is there?

Would the same person be picked if we repeated the experiment? No. Everyone in the world at the start of that experiment had an equal chance of 'winning' but the system guaranteed that one person would be selected. The correct calculation is the odds of picking **any** person, not that particular person, and it was of course certain. If we wanted to calculate the odds of picking any one **particular** person before we started the process, it would be much smaller. It would in fact be 1 in 7.5 billion.

What creationists do is start from what we have now and then attempt to calculate the odds of this particular outcome as it was at some assumed beginning, and this was the intended outcome. But we know that, whatever the probability was, we are here and now and living on this planet. There was no pre–ordination that you and I would be alive today and that you would be reading my book today. That is just the way it turned out, but you would not be reading this now if it had turned out differently. There is nothing special in this situation any more so than there was something special about the person whose name you selected in your mind experiment.

However, there is something incredibly special about you and about every one of us. We are the result of a whole series of processes which, whilst many of them were inevitable, all of them were the result of natural processes. We are the products of the Universe doing chemistry and physics and running according to fundamental descriptive laws. In all the history of the Universe, like each and every one of us, you are unique.

In the next chapters, I'll go through these processes, each of which needed to happen to produce us.

End notes:————————————————

[1] Probability is always calculated as a number between 0 (impossible) and 1 (certain).

3. The Small Matter of Atoms.

In Chapter 1 we looked at the first 10^{-43} seconds of our universe. Now that we have probabilities sorted out in Chapter 2, we can look at what happened next, hopefully free from any suspicion that there might have been some magic involved. We still have some way to go before we have 'stuff' that you and I are familiar with – atoms and molecules and the building blocks of planets and people. The Universe has a lot of spadework to do still.

Remember, about one ten thousandth of a second after the Big Bang, energy was electromagnetic radiation in the form of high-energy photons? At this stage, the Universe's density would have been about a hundred trillion times as dense as water with a temperature of about 1 trillion degrees Celsius. In these extreme conditions, photons would have been moving with enormous energy but would hardly travel any distance before smashing into another photon. The collision force would have been enough to create more elementary particles such as neutrons[1] which would then have been smashed again to produce protons and electrons.

But things were changing rapidly.

When the Universe was just one second old its density had fallen to about 380,000 times that of water, its temperature was down to about ten billion degrees. The number of neutrons now being destroyed exceeded the number being created since photons were still powerful enough to smash them but lacked the energy to create new ones.

At about three minutes old the temperature was down to about a billion degrees (seventy times as hot as the sun) and the rate of change was also slowing.

If the Universe had remained in that state for just a few more minutes, all the neutrons would have decayed to protons and electrons, and that would have been the end. In some of the Universe's histories in the quantum foam, this is indeed what will have happened, but in this particular history, the one in which we live and ask questions about its origins, and maybe in an unknown number of other histories, the temperature cooled enough for some neutrons to stick to protons to form helium nuclei (or alpha particles). Neutrons had become stabilised and your Universe entered its next phase.

We were just a few minutes away from never having existed! Now that is something to ponder; something for you to begin to feel just a bit special about. If the temperature of the baby Universe had cooled just a little more slowly nothing else would have been possible; no galaxies, no stars and planets; no Earth and no you or me. In all the trillions of minutes this Universe has existed, those few are perhaps the most significant for your story because they didn't happen.

Thinks about that for a while!

This phase lasted for several hundred thousand years during which it was still too hot for electrons to stick to the protons and alpha particles to form atoms. Instead, they were free to zig-zag about in the expanding and slowly cooling Universe, interacting with the electromagnetic radiation which still filled the Universe.

This final phase of the Big Bang continued until the next major change which occurred some 300,000 to 500,000 years after the beginning when the Universe had cooled to a mere 6,000 degrees, or about the same temperature as the surface of the Sun. At this temperature, positively charged protons and alpha particles can capture negatively charged electrons to form electrically neutral atoms of hydrogen, helium and lithium.

This was the moment our Universe gave birth to atoms.

There was just enough energy to make lithium out of protons and alpha particles but not enough to make any heavier atoms. That would need nuclear fusion reactions in stars later on. Until then, the Universe had just those three elements and mostly just hydrogen and helium.

The birth of atoms marks the end of the Big Bang and the beginning of atomic matter as we know and understand it. Because these neutrally charged particles hardly interact with electromagnetic radiation, the Universe had become transparent. Photons were now free to stream almost uninterrupted through the Universe and atoms were free to clump together under the influence of their own gravity, undisturbed by photons continually stirring them up and knocking them into different paths so ensuring a chaotic distribution of atoms.

Until this point, the Universe was opaque. Even if we had telescopes powerful enough to look far enough back in time (and looking deeper into space is looking back in time) we could not see past this event.

Although it now had atoms, most of the Universe was still empty space because most of an atom is empty space. Imagine an atom of hydrogen magnified so the single proton in the nucleus is the size of a football on the centre spot in Wembley Stadium. (For those not familiar with Britain, Wembley Stadium is the English National football stadium to the west of Central London). On that scale, the 'orbiting' electron would pass through Durban, South Africa. Everything in between is empty space.

At this point in its history, using classical mathematics, the Universe should have been smooth and without structure – what scientists call amorphous. So, what caused it to be grainy and form clouds of gas, stars and galaxies, etc.? Remember what I said in Chapter 1 about quantum uncertainty and fluctuations? Well, the position of particles in an amorphous Universe would also have been subject to quantum fluctuations and some particles would have been just a little bit closer to others to exert a little more gravitational attraction. A few particles that should have been there would have been missing or in a different place. This meant the gravitational attraction between these particles would not have been exactly equal in all directions, so they would have tended to move together. A quantum universe is not perfectly smooth[2]. As the distance between them decreased, the gravitational attraction increased because it obeys an inverse square law[3], and so they would clump together and have a combined gravity, now attracting other nearby particles.

So clouds of gas could now form in an otherwise amorphous Universe. The Universe was becoming grainy as centres of gravitational attraction were forming; structure was beginning to form under the influence of nothing more complicated than gravity. Order was forming out of chaos and the entire process was inevitable. Nothing that wasn't present or which wasn't creatable from the instant of the Big Bang was needed for this structure to emerge.

And here we have answered another question frequently asked of science, especially by creationists pursuing a religious agenda. How did order come from chaos? The answer is quite simple - gravity.

In fact, the creationist assertion that order cannot come from chaos is demonstrably false. Mathematicians have devised Chaos Theory to explain how it works, and some emergent phenomena can be incredibly structured. Exactly what will emerge and when is not always predictable, indeed Chaos Theory predicts that it will not be predictable other than in very general terms. We know there will be hurricanes and roughly where and when they will form; predicting exactly when, where and how powerful is another matter. To quote a Chaos Theory axiom, a butterfly flapping its wings on a Pacific island can cause a hurricane in the Himalayas.

Examples of structure emerging from chaos could easily fill another book, so the persistence of the belief that it is impossible is something of a mystery. It is not as though no–one has told the creationists who continue to assert it. It is a triumph of dogma over observable reality. All that is required to cause order to emerge from chaos is a directional force of some sort. We have just seen how gravity gives that directional force throughout the Universe. We will meet another directional force when we come to look at biology in later chapters.

Now we have arrived at a point in history where our Universe is making something of itself. Under nothing more that gravity the Universe as we know it is taking shape.

Your story now moves on to the story of stars which are the end-point of collapsing clouds of hydrogen and helium, then the only atoms which exist in our still young Universe.

End notes:————————————————————

[1] Atoms consist of three types of elementary particles – protons and neutrons in the nucleus and electrons outside. There are also other sub–atomic particles of which protons and neutrons are made, but these are the main three, so far as the subject of this book is concerned. Electrons have a negative electrical charge

End notes – continued:

and protons have an equal positive charge. Neutrons are formed from a proton and an electron fused together so they carry no electrical charge. An atomic nucleus has (approximately for larger atoms) an equal number of protons and neutrons and has as many positive charges as there are protons.

One way to think of electrons is as particles 'orbiting' the atomic nucleus but a better way to view them is as a cloud or a series of shells because, being in the quantum domain they can be everywhere at the same time (quantum weirdness!). So, the atomic nucleus is enclosed in a negatively charged cloud spread out around the nucleus. There is the same number of 'orbital' electrons as there are protons in the nucleus, so the atom as a whole is electrically neutral. The reasons things feel solid is because this electron shell repels other negatively charged electron shells.

[2] For a video illustration of an imperfect Universe causing structure to emerge, narrated by Stephen Hawking, see https://youtu.be/DECAorZYErk.

[3] The inverse square law of gravitational attraction states that the gravitational attraction between two bodies is inversely proportional to the square of the distance between them.

4. Stardust

Under gravity, gas clouds fall inwards heating up as they do due to the released gravitational energy. The rate of collapse increases as more atoms of gas are pulled towards the centre and, as increased mass is added, so the gravity increases.

As matter collapses towards a centre of gravity, most of it will have been moving in a different direction to that in which gravity is now pulling them, so it will have some residual 'angular momentum'[1] which must be conserved. Most of this will cancel itself out but any remaining angular momentum must be conserved. This means the collapsing gas cloud will rotate faster and faster as it gets smaller and smaller, so stars and planetary systems, galaxies, even super–clusters of galaxies, all rotate.

To see how this works, watch a skilled ballet dancer or ice–skater doing a pirouette. As they pull their arms in, they spin faster because the angular momentum of the outstretched arms has to be conserved; their arms must travel the same distance round a smaller circle. They slow down again as they throw their arms out for the reverse reason.

Eventually, as this collapsing cloud of hydrogen and helium gets compressed enough, hydrogen atoms get forced together to make helium in a process scientists call nucleosynthesis, and the collapsing gas cloud 'switches on' to becomes a star. The star around which our planet orbits is the sun. It is a fairly ordinary, unremarkable star as stars go.

This process of nuclear fusion releases huge amounts of energy which radiates out from the surface of the sun as electromagnetic radiation (or photons if we think of them as particles) with a very wide spectrum of wavelength. So dense is the centre of a star like the sun that it can take about a million years for a photon to get from the centre to the outside; not because it travels more slowly but because it collides with so many other photons and products of nuclear fusion.

The net outward pressure of these photons being generated in the middle of the star balances the inward pull of gravity. The more the star collapses the more photons are produced and so the greater the outward pressure until a dynamic balance is achieved.

Meanwhile, clusters of stars form into galaxies of billions or trillions of stars which begin to spiral inwards towards a centre of gravity and, as the rate of collapse increases so the rate of rotation also increases until the centrifugal force tending to push bodies outwards equals the gravitational (centripetal) force tending to pull them inward.[2]

Once again, a degree of order is emerging in our Universe from the chaos of unevenly distributed balls of collapsing gas clouds. A new structure is inevitably being imposed on the Universe by the action of a simple force on chaos and that force is once again gravity. A third level of order is also emerging as galaxies form super-cluster.

We do not actually know how many galaxies there are in the Universe. In the 1990s the Hubble Deep Field suggested there were about twenty billion but in 2015 a team from Nottingham University, England, produced figures suggesting there may be ten times as many (Conselice, Wilkinson, Duncan, & Mortlock, 2016). Nor do we know how many stars there are. One estimate is that there are about a hundred billion stars in the Milky Way galaxy alone.

But that is a side issue as far as your story is concerned. Your story is still being written in the centres of these young stars. At the moment we are talking about what are known as first generation stars – the first stars that formed while the Universe was still young.

The precise history of a star will depend on the size of the original gas cloud out of which it was formed. All stars will follow more or less the same process but what happens at the end of their life, and how quickly that end is reached will vary. Let us take a star rather like our own sun in size and about which we know rather a lot and which is known to be a rather average star. Gravity here has produced a temperature of around fifteen million degrees Celsius and hydrogen nuclei (or protons) are being forced together to form helium nuclei and releasing energy as they do so in a huge nuclear fusion reactor.

This process will continue until the star has used up all the hydrogen in its core, where the temperature is hot enough for the fusion reaction to proceed. At that point, with no energy being produced and so nothing pushing the helium outwards, the core will undergo a further collapse, forcing helium atoms to fuse together to form carbon, and the temperature will rise to one hundred million degrees Celsius. The outer layer will swell enormously and, with less radiant energy being produce, the star will dim, so that the sun will become a red giant.

In the case of our sun, this red giant phase will probably swallow up the inner planets, including Earth, and Earth, together with all life on it will become history – just another planet meeting the fate of probably a trillion trillion other planets. In fact, Earth is just about at the outer limit of this inflation so it is not certain that it will be swallowed up but life will long ago have become impossible anyway so it makes little practical difference.

Eventually the helium supply will run out too and, in a final cataclysm, the core will collapse inward one more time and carbon atoms will fuse to form larger atoms such as iron in a final, short-lived burst of activity. This will release energy so violently that it will throw the products of its reaction out into space to form interstellar dust (or stardust) and, together with hydrogen and helium, a new cloud out of which second generation stars will form, but clouds which now containing heavier elements than the hydrogen and helium out of which first generation stars were formed.

Meanwhile, any remains of the core will shrink to a hot, white dwarf, to gradually cool to a black dwarf cinder over maybe the next few million years.

The elements out of which you are yet to be made are, in the early Universe, being manufactured in cataclysmic explosions of collapsing and burned-out stars.

Not all stars have this ending. If they are large enough they will continue to collapse under their own gravity eventually forming a singularity with a gravity field so powerful that not even photons can escape it. They will form black holes and, at that local level, gravity will have regained control and that part of the Universe will again be 'nothing'. This will be a suitable place for another quantum foam of universes to arise, each with its own space and time...

Some stars, significantly larger than the sun may have an ending so violent that they produce a supernova of expanding hot gas that will flare into existence in a matter of weeks and fade over the succeeding centuries.

Some stars may be too large to form red giants and supernovae and too small to become black holes. These will collapse to a density in which all atoms again break down and all electrons and protons are forced together to form neutrons. These are the neutron stars, some of which rotate so rapidly that they produce pulses of radio waves and are known as pulsars.

And some starts will simply go out, not with a bang but a whimper to form white then black dwarfs without the pyrotechnics of their larger relatives.

Your story continues not in a super-massive star destined to be a black hole or a massive star destined to become a spinning ball of super-dense neutrons, or maybe a supernova. Your sun is a second or maybe third generation star. It was formed from a collapsing cloud of gas and interstellar stardust formed in earlier stars and containing the elements from which you will be made; the carbon, nitrogen, sulphur and oxygen in your proteins; the iron in your blood, the phosphorus and calcium in your bones; the sodium and potassium in your cells. You are made of stardust, formed in nuclear fusion reactors and exploding supernovae out of the elementary particles formed out of pure energy during the Big Bang.

End notes:————————————————

[1] The original angular momentum in the Universe came from eddies in spacetime as the Universe expanded early on. This was caused by quantum fluctuations. It means that everything that had that initial momentum has to conserve it, so, unless something intervenes like a collision, everything that develops in that region of space will rotate in the same direction, rather like eddies in water which spins off into smaller eddies, all rotating in the same initial direction. An exception in the sun's planetary system is the planet Uranus which may have been tipped onto its side by a collision in the early stages of the solar system's formation. Earth is tilted, almost certainly due to a collision with another planetoid which resulted in the formation of the Moon.

End notes – continued:

[2] Cosmologists have noticed that the rotational speed of galaxies is far higher than we would expect for the amount of visible matter in them, so there should not be enough mass (mass = gravity) to prevent them flying apart. For this reason, cosmologists think there may be a form of matter, 'dark' matter, which has mass but cannot be seen (because it doesn't reflect or absorb light). The hypothesis is that this is composed of weakly interactive massive particles (WIMPS) that do not interact with other matter. They also think this comes from 'dark energy'.

5. A Home in the Cosmos

Consider again that dot. That's here. That's home. That's us. On it, everyone you love, everyone you know, everyone you ever heard of, every human being who ever lived, lived out their lives. The aggregate of all our joys and sufferings, thousands of confident religions, ideologies and economic doctrines, every hunter and forager, every hero and coward, every creator and destroyer of civilizations, every king and peasant, every young couple in love, every mother and father, every hopeful child, every inventor and explorer, every teacher of morals, every corrupt politician, every superstar, every supreme leader, every saint and sinner in the history of our species, lived there on a mote of dust, suspended in a sunbeam.

Carl Sagan

The late Carl Sagan was inspired to write that by a photograph of Earth seen through one of Saturn's rings by the Voyager space craft. Earth was a barely visible pale blue dot, contained in a single pixel.

But before we talk about human life on Earth, we still have a few chapters of your story to tell. Your story now moves to the formation of this solar system centred on a rather ordinary star in an outer arm of an unremarkable galaxy in a cosmos, containing perhaps a 200 billion galaxies each with maybe a 100 billion stars.

It begins about 4 billion years ago by which time the Universe was about 10 billion years old. It happened in a spiral galaxy we call the Milky Way because it looks like a whitish cloud of stars across the sky when seen from the surface of Earth when we look toward the centre of the galaxy.

Within the Milky Way galaxy, in one of its spiral arms, a rotating cloud of gas and stardust collapsed to form a new star – our sun – just as it had done trillions of times before in this and other galaxies. The heavier elements, formed in earlier generation stars, would have been thrown outwards again as the nuclear fusion machine switched on. These would have formed a disc of debris consisting of atoms and molecules of heavier elements – a so-called accretion disc. Matter in this accretion disc would have begun to condense around stable centres orbiting the sun as planets began to form.

Radiant energy from the sun would have driven the lighter gasses further out so that the outer planets are gas giants consisting mostly of gases with a rocky core and the inner planets are rocky. The accretion process was by no means a steady gradual process but, as clumps of matter grew larger, so the impacts of accretion grew larger. On bodies like the Moon and Mars with their stable outer layers, we can see the craters which resulted from the impacts of large bodies such as asteroids.

One collision involving Earth early in the life of the solar system was with a very large object, possibly a small planet, which tilted Earth on its axis of rotation and threw enough material out of Earth's gravity to give rise to a small accretion disc of its own. The relatively large Moon was formed out of this debris. For all practical purposes the Earth and Moon form a twin planet, made out of the remains of these two earlier planets.

This event had huge significance for your story. The tilt of Earth's axis has produced seasons, and the close proximity of another relatively large body has produced tides. It is probable that both these contribute to Earth being a suitable place on which life could exist, and it is certainly a major part of why life on Earth is as it is today. It is very probably the reason living things could move out onto land as we shall see later.

Earth has an almost solid core containing mostly iron with some nickel, giving it a magnetic field. Around this inner core are molten layers comprising the outer core and the upper and lower mantle on which sits the thin oceanic crust and the thicker continental crust. The continental crust is what the major land masses are comprised of.

Because the mantle and outer core are basically viscous liquids, and because of the currents in them caused by the rotation of the Earth, there is pull on the continental crust causing it to break into plates that move relative to one another – plate tectonics. As they move, the plates can slide under one another causing subduction trenches in the oceanic crust near the edges of continents, or push one–another up into mountains. The Himalayas is the result of the Indian plate pushing up against Asia; the Alps are due to Italy being pushed by

Africa up into Europe. The Atlas Mountains in Africa, the Pyrenees between Spain and France and the Chiltern Hills, Berkshire Downs and Kent Weald in England are all the result of Africa pushing the Iberian plate into Europe.

So, Earth is a dynamic, constantly changing place where continents move, form larger land masses and break up again, carrying their passengers into higher or lower latitudes , pushing mountain barriers up to isolate them or create land bridges, eroding them down again, and recycling its surface over millions of years. Continents are eroded by frost and water and washed into the oceans as sediment by rivers that form river valleys, water meadows, marshes and deltas.

Meanwhile cracks in the oceanic crust allow the mantle to well up to form the deep ocean 'black smokers' or geothermal vents, the mid–ocean ridges and occasionally volcanic islands such as Iceland, the Hawaiian and Galapagos Islands. In places, especially near the junctions of tectonic plates, volcanoes and earthquakes occur. It is no coincidence that places with volcanoes, like Japan and Italy also have earthquakes.

As the continents move, ocean currents change, sometimes bringing warm water from the tropics far enough north or south to melt the polar ice, sometime a little; sometime a lot. At other times, these currents fail, and the ice extends again. Sometimes volcanic dust dims the sun for years and occasionally, a large asteroid comes burning through the atmosphere to devastate the planet.

On this dynamic little planet something happened. It might well have happened **because** Earth is such a dynamic place with such a range of changing conditions on both a large and small scale. The most likely place it happened, according to current consensus, is around the deep–ocean black smokers. It was because of this that we are here and not just a disorganised collection of the elements we are made from.

There are still a few chapters of your story to tell, but first, let's talk first about entropy, the Laws of Thermodynamics and one or two other things about physics.

6. Getting Physical

We are about to consider life, what it is and why there is such a thing on Earth but first we need to understand a few basic scientific principles that many people think they understand but do not. There is a rich seam of misunderstanding and ignorance here for creationists to exploit, so let's dispel the myths.

There are three laws known collectively as the Laws of Thermodynamics, which deal with energy.

The First Law of Thermodynamics:
The amount of energy in an isolated system is constant. Put another way, this means that energy can neither be created nor destroyed; it can only be transformed into another form. Until Einstein showed the relationship between energy and mass with $E = mc^2$, science used to say that matter or mass can neither be created nor destroyed. We now know that matter and energy are interchangeable. You might hear this called the Law of Conservation of Energy.

The Second Law of Thermodynamics:
The total entropy of an isolated system can only increase over time. The important words here are 'isolated system'. Entropy can decrease locally and can decrease in an open system if energy is supplied to the system. By 'entropy' scientists mean the amount of disorder. Systems tend to become disordered unless energy is used to keep them orderly. There is nothing really special about this descriptive law; it is an inevitable consequence of statistical probabilities because there are more ways to be disordered than ordered.

Consider a bunch of a dozen or so helium–filled balloons, all held in your hand by their strings. You let them go and they float into the air. Forgetting for the moment the effects of any wind, what will happen to the balloons as they rise into the air? Will they stay bunched together (ordered), or move apart (become less ordered)? They will tend to move apart simply because there are more places for them to randomly move away from one another than there are for them to move closer. Locally, two or three balloons might well become more ordered by moving closer together but the system as a whole will always tend to become less ordered.

I don't know about you, but my house always seems to need me to use some energy to keep it tidy. No matter how I throw my cloths at the wardrobe they always seem to find more ways to be disordered than they do to hang themselves up neatly!

Earth is not a closed system! It seems to have escaped many creationists notice but we have a sun in the sky which is pumping energy out, some of which reaches Earth. The entropy in the sun is increasing but on Earth that energy can cause a decrease in entropy, so long as the total entropy doesn't decrease. In fact, the notion of a truly isolated system is a theoretical one since the only truly isolated system is the entire Universe.

But there is nothing in the Second Law that says order cannot emerge from chaos. Consider the Bridge hands we met earlier in Chapter 2. What would be the probability of starting with a randomly shuffled pack and dealing four 'perfect' hands, each containing all the cards in a single suit? In fact, it is exactly the same as dealing any other four hands. We define what is 'ordered' and what isn't and clearly those hands could be more or less ordered on that definition.

In London, you can expect to see maybe one black cab for every 20–30 private cars, and maybe one red double–decker bus for every 20–30 black cabs. Now, if we always saw private cars, black cabs and red busses evenly spaced out, we would regard that as a highly ordered system, yet anything else also has a degree of order in it. We can expect to frequently see three red busses together and maybe four or five black cabs in succession. This arrangement is more disordered than a strict sequence according to relative numbers, yet we would regard three busses together or five black cabs together as local order. The random movement of traffic has produced local order in the random sequence of traffic. In fact, we would suspect something was amiss if this were not the case. We expect clumping and local order. It is not forbidden at all by the Second Law. Statistical probability predicts and demands it.

The Third Law of Thermodynamics:
This law states that the entropy of a perfect crystal at absolute zero is exactly equal to zero. It need not concern us very much, except to note that this perfect order with zero entropy is produced by the operation of the descriptive laws of physics.

But what does this have to do with life?

Let's look now at what 'life' really is. It can be fun to ask a creationist to define 'life' just after they have been telling you that science cannot explain how it came about. Invariably, you'll find that they do not know what they mean by 'life' and their religion doesn't define it very well either, if at all, so they wouldn't recognise it if a scientists showed it them. They have been quoting a badly thought-out dogma, not a reasoned scientific position with scientific definitions.

Often, it seems they think 'life' is some magical ingredient like a spark or a special form of energy, or even a tiny imp. You'll never get them to even say whether they think 'life' is a substance, a process, or something else entirely.

The following is based on a blog post I wrote some years ago to try to explain this (Rubicondior, Rosa, 2013):

> *We all use words like 'alive', 'living', 'lively', 'life' without any real thought as to their meaning. We all know what we mean when we talk about a new life, do we not? We mean a new instance of life; a new example of it, as though some new 'life' has entered the Universe.*
>
> *So, what is this stuff called life? Let's see if we can define 'living' as it applies to 'non-living' things:*
>
> ***Living things grow and repair***. *Well, yes, they do, to an extent. But single-celled organisms do not repair in the sense that a human can grow some new skin to heal a wound or can grow new bone to repair a fracture. They may be able to repair some damaged membranes, but growth is limited to splitting into two smaller versions and then growing to full size to repeat the cycle.*

Living things reproduce. *Certainly, some do but many do not, yet are none-the-less alive.* *Yes, their body cells may be reproducing as part of growth and repair but reproduction of new individuals is only a function carried out by **some** living individual, which are no less living for not reproducing.*

Living things breathe. *It depends on what you mean by 'breathe'. Very few organisms actually shift air in and out of lungs like mammals do. If you mean they have a process of respiration - that is, they take in oxygen and use it for metabolic processes and give off carbon dioxide as a waste product, then many 'living' things certainly do that, yet not all. Anaerobic organisms, like many bacteria do not use oxygen; in fact, it is toxic to them. But yes, in the very general sense of exchanging chemicals with their surroundings, respiration can be said to be a characteristic of living things.*

Living things move. *Many do; many do not. A sedentary organism is no less alive for not moving.*

Living things metabolise. *Yes! Finally, we have something that is common to all living things! All living things have chemical processes going on inside them. These processes essentially overcome the tendency towards disorder. We call disorder 'entropy'. This tends to increase and can only be reduced locally by increasing it elsewhere. An organism is essentially an entropy reduction machine, using an energy source (an increase in entropy) to reduce local entropy (metabolism).*

So, we can say that 'life' is entropy reduction at its basic level, can't we?

What then of the idea of 'new life'? Let's look briefly at how a new multicellular individual, like a mammalian baby, gets made:

Special cells called eggs, or ova, are produced by the female from organs called ovaries. Males produce other special cells called sperm cells from organs called testes. Inside the female's body (in a mammal) a sperm cell enters the ova and fertilises it, so a new cell, made from both the male and the female, is produced and a new individual begins to develop from it.

Yet both the ova and the sperm were alive during all stages in this process. No 'new' life was created at all. All that happened was, the existing 'life' continued. In other words, the metabolism continued; the entropy reduction machine continued to overcome the tendency of entropy to increase locally. There was no new life; only a new entropy reduction machine.

So, what then is 'life'?

Life is the local reduction of entropy using the energy released by an increase in it elsewhere. We call this 'metabolism'. Life is what we call the biochemical reactions inside living things. There is no special ingredient which makes the chemicals inside a living organism different in some way to other chemicals. The chemicals inside living things are obeying the same basic chemical laws as are the rocks in the ground and the gasses in the air. There is no magic ingredient that science cannot explain. Religious people who insist science should be able to explain where 'life' came from are asking science to explain a superstition for which there is not a single piece of evidence. Of course, science can't explain it. Nor can science explain fairies.

And when life ceases, the metabolism stops; or rather when the metabolism stops, life ceases.

So, the question that science **can** address is how did this entropy management system arise? What chemical and physical processes can produce it? This is the abiogenesis question and it has nothing to do with evolution because evolution is about what happened **after** 'life' got started; how it diversified, changed and developed.

The honest answer to this question is that science has not completely answered it yet, but scientists have made considerable progress in explaining some of the possible mechanisms (there need not have been just one).

As I mentioned in the previous chapter, one of the best candidates for where it happened (and it need only have happened once!) is a deep–ocean black smoker or geothermal vents. These vents form a unique ecosystem with species not found anywhere else. They also rely on an energy source other than the sun – geothermal heat and instead of oxygen they use sulphur. Heat and sulphur they have in abundance around the black smokers.

The bacteria around these geothermal vents are known as chemosynthetic or chemoautotrophic bacteria and are able to synthesise organic molecules from primarily hydrogen sulphide. The entire ecosystem depends on these at the base of the food chain so these vents have a density of organisms some 10,000 to 100,000 times higher than would be if they relied on the 'snow' or organic debris falling down to the sea floor from above.

Lots of experiments have been done to show how simple organic molecules can be produced from a mixture of various precursor chemicals to produce the building blocks of proteins, amino acids and the basic elements of RNA[1]. Most importantly, some short lengths of RNA have been shown to be able to assemble these basis elements and produce copies of themself.

Incidentally, it is amusing to watch creationists trying to make the best of this science because it neatly illustrates the disingenuous nature of their approach to science. On the one hand, they will claim that since scientists have not created 'life' in a laboratory, this shows it cannot be done and so couldn't have occurred naturally. On the other hand, when they are shown that scientists have gone a considerable way towards producing self–replicating systems, suddenly this proves it cannot be done naturally and requires intelligence. The goal–posts are moved.

It doesn't require intelligence to make the chemical and physical reactions happen of course, any more so than it takes intelligence to make hydrochloric acid and sodium hydroxide react to make sodium chloride (table salt) and water. All the scientists do is create the right conditions and supply the right reactants, all of which could and do occur in nature.

'Life' is nothing more than chemistry, specifically, chemistry which uses energy to resist entropy. 'Life' is entropy management.

Getting Physical

End notes:————————————————————

[1] Ribonucleic acid. RNA differs from DNA in having a ribose sugar instead of a deoxyribose sugar. It is thought that the simplest self–replicating systems (can we call that 'life'?) were based on RNA like many viruses today are. The 'code' in DNA is 'read' to produce a strand or RNA which then acts as the template for making a protein, so the basic metabolic processes can be seen as RNA–based with DNA simply being RNA's data store.

7. Living the Good Life

There are several hypotheses about how it happened but however it happened, about 3.5 billion years ago, on this little planet orbiting a very ordinary sun situated on an outer arm of a very ordinary galaxy, something quite extraordinary happened. It might well have happened on other planets in other galaxies maybe a trillion times but, as the distances between galaxies and even across galaxies are so vast, we may never find out. It might have happened more than once on Earth and it might be that several different solutions got together by chance to produce a system that was much better than either one alone.

Here on our tiny bright blue dot, a replicator arose. This replicator, or one very much like it, is your earliest ancestor. There is an unbroken line between this first replicator and you!

A replicator is a chemical which can produce copies of itself. We do not know exactly what the first replicator was; one theory is that it may have been crystalline structures in clay; another is that it may have been ribonucleic acid (RNA). It is generally agreed that RNA came into it fairly early on, maybe from the outset, maybe riding on the back of some other chemical in an evolutionary process in which those replicators better at replicating will produce more descendants. There are still some organisms like viruses which depend on RNA and RNA is certainly still involved even where the main genetic material is deoxyribonucleic acid (DNA). However, this replicator started, it seems to have fairly quickly evolved into a DNA-based one.

This is where creationists try to play the big scary numbers trick again by acting as though science claims the 'first cell' came into existence fully formed in a single spontaneous event. Then they try to put some guessed and very tiny probabilities together to make an all–but impossible probability, declare it to have been impossible and proclaim 'God did it!' with no hint of a mechanism by which this happened nor the slightest piece of evidence to support the idea, nor any *a priori* evidence that such a god actually exists. Wilful misrepresentation of science and ignorant incredulity presented as scientific data is followed up with a false dichotomy fallacy.

No serious biologist working in the field of abiogenesis has claimed that a fully–functional cell spontaneously self–assembled somewhere. No serious biologist every proposed as an explanation for abiogenesis that a complex modern cell suddenly assembled itself out of some sort of soup or even, as is sometime bizarrely claimed by creationists, from rocks. No scientist would be that naïve. Creationist charlatans however know that many of their willing dupes will be and even if they're not, will gladly pretend to themselves to be, and will gladly swallow the assumption that somehow scientists are either mad or stupid, or probably both.

There could have been billions of simple self–replicating systems, competing for limited resources with the most efficient making the most copies. A billion to one chance only needs to occur once in every cycle of self–replication. With billions of systems going through tens, hundreds, maybe thousands of cycles a day in the right conditions, the billion to one chance comes up regularly. The chances of you winning the National Lottery, assuming you buy a ticket, are tiny and yet most weeks someone wins the National Lottery because millions of people buy tickets.

It could have taken tens or hundreds of millions of years before anything resembling a cell as we know it gradually evolved.

Each of your cells, with the exception of your red blood cells, contains DNA which contains all the instructions for making you and making the protein enzymes which control your body's metabolism. DNA is organised into genes which carry instructions. Without going into detail (whole books are devoted to how DNA works as a store for information and how it is organised into genes, but the details are very well understood), these genes are the basic building blocks of life and are the replicators which carry your information from your parents, through you and into your children. Your story is the story of your genes which, over billions of years have built survival machines for replicating themselves through time. You are a gene survival machine with a body superbly fashioned to survive and reproduce and so pass on your genes to the next generation.

Replicators are particularly good at replicating but occasionally they produce slightly different copies of themselves, and these slight differences give rise to slight variations between individuals. Some of these will make the individual less fitted to survive and some will make it more fitted; the vast majority however will make no noticeable difference as they take place in the accumulated redundant DNA which all organisms contain.

Whether that variation will be advantageous, detrimental or will make no difference at all will depend on the environment in which it finds itself. The environment will tend to select for individuals carrying genes which make it more able to survive and against those less able. Those which tend to survive will tend to produce more descendants, in that environment, and those descendants will inherit the advantageous genes.

The environment gives meaning to the information in the genome. Take for example a hypothetical bacterium that by a chance mutation found it could digest man–made vulcanized rubber. Vulcanizes rubber is made much stronger than the natural latex by creating sulphur–sulphur bonds across the long–chain natural molecules. Millions of tons of vulcanized rubber a year are turned into fine dust by car tyres wearing out on roads all over the world.

To this hypothetical bacterium, would this ability be beneficial, detrimental or neutral? It would of course be beneficial because of all the vulcanized rubber in the environment. Two hundred years ago it would have been at best neutral. If it inhibited the bacterium digesting its 'normal' food, it would have been detrimental. The information in the genome only has meaning in the environmental context. With a constantly changing environment, the meaning of the information in the genome is also constantly changing. Do not be fooled by creationists constantly gibbering about information in DNA as though that's a fixed thing.

Natural selection acts like a sieve at each generation tending to filter out those characteristics which act against survival and allowing through those which enhance it. In this way the gene pool of a species tends to adapt to its local environment and, where the gene pool is split amongst several different environments, tends to produce diversity. Diversity will eventually lead to speciation if separation is maintained for long enough.

Evolution by natural selection is not a complicated process to understand; in fact, it is quite simple. Thomas Huxley, on reading the first edition of Charles Darwin's, 'Origin of Species' is reported to have said, "How stupid! How stupid not to have thought of it oneself!" It really is so simple it takes great intellectual creativity and mental gymnastics to not understand it. It is far easier to understand and explain than gravity. There are only three things needed for evolution to happen:

1. Inheritance of physical characteristics.

2. Imperfect reproduction of those characteristics to give variation.

3. An environment which favours some variations over others making it more likely they will be passed on to the next generation.

Does anyone seriously doubt any of these? If so, which?

Inheritance of physical characteristics?
Haven't you noticed how children usually look quite a lot like their parents? Haven't you noticed how fish tend to have fish offspring and birds tend to have offspring which look a lot like themselves?

Imperfect reproduction of those characteristics?
Haven't you ever noticed how you can tell the difference between individuals of many species? That some individuals have distinct colours to others or different markings? That some individuals are bigger, smaller, faster, fatter, or thinner?

An environment which favours some variations over others?
Do you think an animal which gets eaten is going to have more offspring on average than one which doesn't? Do you think an animal which finds food more easily is going to produce fewer offspring than one which cannot get enough to eat? Or do you think this won't make any difference? How about an individual who finds a mate more easily than another or who rears its young more successfully? Which do you think is going to produce the most descendants on average?

So, if you believe evolution doesn't happen, you have to tell me what is impossible in any of these steps. Simply repeating a dogma in the face of the logic won't work. You have to say why.

If you cannot tell me that, you have to tell me why these three simple steps do not lead to more of the physical characteristics which were favoured by the environment, being present in the next generation. You have to say why the ratio of those different variations won't change over time in such a way as to make the individuals carrying them better at surviving and producing offspring in that particular environment?

Because, if you cannot tell me that, you do not disagree with evolution even though you wish you did and even though you might have a book which says it didn't happen. In fact, if you cannot tell me why this is wrong you're agreeing that evolution not only happens but that evolution must happen if those three simple steps are present and without any magic.

In other words, if you accept that these three simple steps occur, then you believe in evolution just as completely as does any evolutionary biologist. If you **do not** accept them then you are indulging in an extraordinary degree of denialism and self–deception.

That explains why evolutionary change occurs due to environmental change and we saw in Chapter 5 how Earth's many different environments are constantly changing, so there is always pressure on life in general to continue to change. It is not aiming for some endpoint, merely being forced by the changing environment to adapt to it. Those that fail become extinct, at least locally.

So, natural selection by the environment is the directional force I alluded to in Chapter 3. The direction is towards improved ability to replicate in that environment and the 'chaos' that order emerges from under the influence of that force, is variation and random copying errors in the replication process. Biodiversity and the order we see in living taxa is an emergent property under the directional force of natural selection; nothing more and nothing less.

But that doesn't entirely explain why different species are produced by the evolutionary process because, if a species can have a range of variations what limits this range? Why can all forms of life not interbreed?

It is a common misconception that evolution is about how species evolve and that somehow the purpose of evolution is to produce new species. There is no such purpose; speciation is merely an incidental outcome of evolution but there are some biological reasons why it occurs and especially why it can occur quite rapidly in the right situation. 'Rapid', in evolutionary terms means in tens or thousands of years, not millions, although there are some examples of it happening even more quickly than that and some of it never having completed after many hundreds of thousands of years. We will meet some of these examples when we look at the evolution of modern humans.

I've tried to explain this is several blog posts over the years. The following are two examples explaining different scenarios, first, European finches (Rubicondior, Rosa, 2011).

> *So why does speciation occur at all?*
>
> *In the standard model, the first essential is that a group of individuals becomes isolated from the main population for long enough for gradual changes to accumulate in response to local environmental factors. These factors may be predation, success at finding food, breeding success, etc. The main population will meanwhile be changing in its own way in response to its own local environment, or not, if the environment is stable. Both populations will now be set on their own evolutionary trajectory, independent of one–another.*
>
> *Eventually these differences may build up in each population so that, if ever they DO come back into contact again the two populations' genetic make-up will be such that they physically cannot interbreed successfully to produce fertile offspring, even if they can still successfully mate. This is the case with donkeys and horses, lions and tigers and many species of plant.*
>
> *In this model, speciation is merely a passive, incidental result of gradual evolution. There is another model however, in which speciation is itself driven by natural selections. (I'll use European finches to illustrate this but I could equally have chosen almost anything else; insects, reptiles, plants, fish or frogs, etc.)*
>
> *Consider Europe either side of the last ice age. Northern Europe, the Alps to the north of Italy and the Pyrenees between Spain and France*

were all heavily glaciated, driving many species south into Spain, Italy and the Balkans and effectively isolating them there with impassable ice sheets.

Now take a species of finch, adapted to live in Northern Europe with a generalised bill for eating a variety of seeds. This would have been pushed south to form several isolated populations. Each would have evolved and adapted to best use the evolving and changing plant population.

One species in, say, Spain, may have evolved a slender beak for picking seeds from thistles and other wild flowers, the more successful ones passing these beaks on to their offspring. The other population in, say, Italy, may have evolved stouter beaks for cracking harder seeds, also passing these on to their descendants. The two populations would be diversifying according to local conditions.

And of course, the plants would have been evolving under this selection pressure with those which avoided having their seeds eaten being favoured by the environment. Finches and plants would be in an evolutionary arms race, pushing the finches into an increasingly specialised form.

Now move on to the end of the ice age when the Alps and Pyrenees became free enough of ice for the finches to return, together with their food plants, into an increasingly temperate Northern Europe as the ice retreated:

Suppose these finches had not been isolated for long enough to make interbreeding impossible; after all there was then no evolutionary pressure to cause that to happen. If it had, it would have been through chance alone. What type of beak would their offspring inherit? They would probably inherit an intermediate beak, but an intermediate beak which was no use for either of the favoured food plants of its parents. To all intents and purposes, they would be handicapped and incapable of feeding or capable of feeding only with difficulty.

These would be rapidly removed from the gene-pool. Interbreeding would be hugely wasteful as the result of all that effort would be a lost brood, or, at best, a brood of individuals with a greatly reduced chance of themselves producing offspring. Anything which favoured non-interbreeding between the two forms would now be highly favourable. There would now be strong selection pressure for barriers to hybridization and reproductive isolation to evolve.

Changes in plumage combined with display mating rituals, territorial and mating song, and, especially, female sex selection would all be favoured.

Gene pool isolation would be reinforced now, not by geography but by any other means available. A process of speciation which began casually and incidentally in, and because of, isolation, would now be accelerated paradoxically by a lack of the very isolation which initiated it.

*So, we have lots of different finches in Europe, each with its own plumage, song, mating rituals and food plants, many of which are actually **still** capable of interbreeding successfully, and do so in captivity, but which rarely do in the wild. Incidentally, notice if you will that never would there have been an 'intermediate form' between the two new species such that creationists demand to see as evidence of evolution, because one did not evolve into the other; they both evolved from a common ancestor.*

Speciation has occurred because it was in the 'interests' of both gene pools to speciate. An incidental yet inevitable result of evolution and an undirected, yet highly directional, process of natural selection acting as though it were driven by the needs of genes to replicate through time.

This next example shows how environmental change alone can produce evolutionary diversification (Rubicondior, Rosa, 2012):

Consider first a hypothetical situation in which no species, sub-species or variety had ever gone extinct, so every twig on every branch of the tree of life had living representatives. Apart from the mammoth task involved in catching and classifying the resultant billions of 'species', how could we possibly classify individuals and place them into a specific taxon?

It would of course be impossible because in reality we would have a mass of living things with no sharp delineation between groups of individuals having characteristics in common and distinct from all others. The dividing lines would be so blurred that we could not easily identify any boundaries. It is only because of regular extinctions and gradual evolution leading an archaic form to change into a modern form across the entire population range, that we can see living things as forming distinct taxa in the first place.

We can see the present, but the past is invisible to us directly and can only be inferred from evidence like genetics, morphology, physiology, embryology, with the occasional fossil lending support. It is as though the tree of life were growing underground, with only the very tips of the surviving twigs showing above the surface. And of course, that's exactly where the few fossilised remnants of the tree are - in the ground.

The problem with this tree of life model is its simplicity. It is, after all, only a model. It tends to make us lose sight of the fact that a 'species', represented by a single twig on this tree, is not a single entity but tens or hundreds of thousands, even many millions of individuals, all of which are actually little twiglets in their own right. At any one point in time, past, present or future, any group of individuals has the potential to be the start of a new twig on the tree. All that is needed is for them to become genetically isolated in some way and then to evolve in a direction which makes them eventually incapable of breeding successfully with descendants of the parent species.

One problem with this over-simplified model is it is easy to fool simple people with it. For example, when creationist pseudo-scientists reassure their credulous customers that no one has ever seen a new species arise from an existing one, it looks like a rational argument against the Theory of Evolution because they think of a 'species' as a single entity which the Theory of Evolution says suddenly 'evolves' into a new species in a single event. This gives rise to the 'no one has ever seen a monkey give birth to a human' syndrome. Of course, the Theory of Evolution only bears a superficial resemblance to this creationist parody and says nothing of the sort.

In a very few instances, a new species can arise because of a single mutation, such as the example I gave in The Good Shepherd's Purse Is Bad News For Creationists (Rubicondior, Rosa, 2012). *Occasionally in plants, it can arise by hybridization between related species, as with bread wheat. But, for the vast majority of speciations, the process is a slow, gradual one in which a group of individuals become reproductively (i.e. genetically) isolated for a prolonged period of time.*

But let's think about those situations where speciation is, or could have been, a single event such as a mutation or a hybridisation. Even if you witnessed it would you have knowingly seen a new species arise? Would you regard a single individual as a new species, or just a mutant? It is only after a population has been established, several, maybe many, generations later that anyone is going to know that Earth has a new species. By then, the precise location and the precise founder individual will be unknown and unknowable.

I'll illustrate this with a hypothetical example. Let's take a species of monkey living in a large rain forest spread over several thousand square miles. These monkeys will be able to move freely across their entire range so that genes can flow and spread throughout the entire gene-pool and any evolution due to selective environmental pressures will occur in the entire population.

But, gradually, due to climate change or continental drift, or maybe a change in ocean currents, the forest begins to get drier and turn into grasslands, with trees surviving only close to rivers. In other words, the monkey population is broken up into isolated groups which can no longer interbreed because they simply do not come into contact any more. Each group will be free to evolve according to the local conditions in its woodland. Eventually, maybe after a few hundred thousand years, maybe a million or two, these groups may evolve to the point where they not only look different to each other but may not be able to interbreed even if they do meet up.

So where and what was the 'speciation event'? At what point in the process could an observer say, "Hey! I've just seen speciation occur! It happened when..." In fact, we only know that speciation has occurred retrospectively because, according to our rules of taxonomy, failure to interbreed means they are now different species. Maybe if we had been able to examine them a hundred thousand years ago, we might have found that they could still interbreed. Maybe we would have found an incompletely speciated 'ring species'.

There was no sudden emergence of a new species; no sudden branching of the tree of life; no mutation which brought a new species into being and no 'macro-evolution' event. There was no event which creation pseudo-scientists proclaim to be impossible and which they claim has never been seen. It was just a slow accumulation of difference, directed by natural selection with each group doing nothing but struggling to survive and reproduce with the ones which left the most descendant contributing the most genes to the gene-pool.

Now, take the same scenario, only this time the climate changed again after a few tens of thousands of years and the isolated scattered groups could once again mix freely. But this time maybe they had not diverged sufficiently to prevent interbreeding, or maybe one group now had a significant advantage over the others. In these cases, the group with the genes which gave them greater success would come to dominate and possibly replace the others.

Is this speciation? Is this the point at which we can say a new species arose and the 'archaic' form went extinct? Or is this merely evolution of the entire species? Were those groups isolated for a few thousand years new twigs on the monkey branch of the tree of life, or were they merely groups of individuals with the potential to become new species, but which never quite made it? Certainly, the day they came back into contact, nothing happened to their genes. It was not a change on their part which caused them to re-establish contact. It was the environment which changed. And of course, we can never know whether they could interbreed or not, so we will never know if they were ever a distinct species according to our rules of taxonomy. More importantly, what was there an observer could have seen as a speciation event? Again, there was nothing because it was not an event, it was the result of a process.

And it actually makes not one jot of difference.

To a palaeontologist, examining fossils from across this period, if he/she were fortunate enough to find any, it might look like there were different species at one point, and then all but one of them suddenly went extinct, leaving just the present species. Compressing geological time into the geological column where a few thousand years can be a few inches, it might look like climate change caused extinctions with only one 'species' of this group of monkeys surviving it.

A creationist 'scientist' would of course chortle at the lack of transitional fossils and probably write another book about how this has destroyed 'Darwinism' and proved it was the Christian god who made the monkey 'kind' 6000 years ago.

(I'll resist the temptation to point out that someone is making a monkey out of someone - oh! I just did)

To nature, all that has happened is that living things have evolved according to the pressures in their environments. It is the environments which have changed and life has adapted to it. Nature has not the slightest interest in making new species. They arise only incidentally in the process of evolution and usually nobody and nothing pays it the slightest attention because nothing remarkable, or even recognisable, happens.

And we only know about it when it is done and dusted.

Speciation is rarely a single event. It is a process spread over a prolonged period, and we do not know it has happened until well after

the event. Even if, as in the rare case of a single mutation, or a
chance hybridisation, we would not regard it as speciation unless,
many years later, we find it has produced successful descendants.
'Failure' to observe a specific event says nothing about the validity of
the Theory of Evolution because the Theory of Evolution tells us not
to expect one.

Contrary to creationist claims, the Theory of Evolution by Natural Selection is probably the best supported theory in the whole of science. No serious biologists now spend time trying to prove the theory, and especially not trying to disprove it. It is the foundation of all biomedical science and is the grand unifying theory of biology without which much of biology would not make sense.

Creationists frequently try to comfort themselves and their customers by declaring it to be a theory in crisis and about to be overthrown and replaced by one which includes magic and explains nothing – Intelligent Design. Intelligent Design is the pseudo–scientific creation of the Discovery Institute, a fundamentalist Christian front organization (Discovery Institute, Undated) which seeks to have Bible literalism taught in American public schools despite the constitutional prohibition on promotion of religion by American federal or state government or its agencies.

In the Kitzmiller vs Dover District School Board trial, the judge found that many of the fundamentalist Christians on the school board had lied in court about their religious beliefs, claiming they had no religious interest in having Intelligent Design on the science curricula in their schools when in fact they were creationists and leading members of a fundamentalist church that advocated for Young Earth Creationism and a literal, six–day creation interpretation of the Bible. They had got themselves onto the school board specifically to get their views promoted to school children at public expense. The star witness for the Dover School Board, a leading Intelligent Design advocate, Michael J. Behe, was forced to admit under oath that:

There are no peer reviewed articles by anyone advocating for
intelligent design supported by pertinent experiments or calculations

which provide detailed rigorous accounts of how intelligent design of any biological system occurred. (Court Transcript, 2005)

Michael J. Behe

In other words, Intelligent Design is not science but narrowly sectarian fundamentalist religion in disguise.

We have seen now how evolution by natural selection works. I'll now run briefly through the way genetic information is stored, and particularly why there are differences for natural selection to work on. I do not intend to go too deeply into this subject because there are entire books dedicated to DNA, genetics, etc.

First a little technical details about inheritance and genes, because what we are really talking about in evolution is evolution of genes.

Genes are carried on the DNA that we get from our parents. Humans have 23 pairs of chromosomes (packets of DNA) giving 46 in all. Two of these are the 'sex' chromosomes, given the letters X and Y. Females have two X chromosomes (XX) and males have one of each (XY). With all the other 22 pairs both sexes have two of each.

Genes are normally thought of as discrete sections of DNA that contain the code for making a specific protein[1]. Proteins have lots of different roles in the body from structural proteins like collagen and keratin to enzymes that control chemical processes.

So, we inherit our distinctive characteristics from our parents in our genes. But, as we know, one of our parents might have one eye colour, hair colour, skin colour, etc., and the other parent might have different coloured eyes, hair and/or skin. They have different genes for these characteristics. These genes are in the same position on the same Chromosome, but they are slightly different. These differences are called alleles.

If you look at people in the street in, say, New York, you will see a complete range of different people of different heights, different hair types and colour, different skin tones, differently shaped noses and lips. If you could look deep inside you would see thousands, maybe millions of slight differences between people. You would find different alleles for blood groups, for lactose intolerance, for the ability to smell certain smells; for a whole range of small internal differences that might or might not affect health, etc.

If you look at people in a street in Delhi, Beijing, Cuzco or Kinshasa you will see a different range of types, some less variable than in New York and some more so. Most of these differences would obviously be the more superficial and more noticeable ones but if you could look deeper inside again, you would find many of the different alleles that are deep inside New Yorkers, but they would differ in frequency across the respective populations.

When biologists talk about evolution, all they mean by the term is change in allele frequency over time. But what causes this change? The local environment tends to favour some alleles over others so those which are more beneficial to the individual with them will help that individual survive and have more children. Those that are less well favoured or which are deleterious will mean the person will have fewer children on average.

Different populations have different frequencies of the different alleles. Different blood groups for example, have different frequencies in different populations. The reason for these differences in different populations is, because before travel made intermixing easier and more likely, for most of our history humans lived in lots of more or less separate groups in different environments in different parts of the world. Those environments selected for different sets of alleles as the populations adapted to local conditions. Of course, we were never entirely isolated so genes could 'flow' from one population to another and populations never remained isolated for long enough to evolve into different species or subspecies, at least so far as modern humans are concerned. It was not always so for our ancestors, as we will see later.

Before I move on to the evolution of modern humans, there is a much earlier part of your story still to tell; the story of how simple self–replicating molecules eventually gave rise to the ancestor of the human family that founded our small branch of the tree of life, the hominids.

End notes:————————————————

[1] The DNA 'code' comes from the fact that DNA is built from four different nucleotides. Each nucleotide consists of a sugar (deoxyribose), a phosphate group and a nitrogen base. There are four different nitrogen bases giving four different nucleotides: cytosine (C), guanine (G), thymine (T) and adenine (A). RNA is similar except that thymine is replaced by a different base, uracil (U) and the sugar is ribose. These bases are joined in chains and two chains are arranged in the 'double helix' with cross–linkages between like the rungs of a ladder. The side chains are produced by the sugar and phosphate groups and the cross–linkages by the bases. Adenine always links across to thymine and guanine to cytosine. A single chromosome is a massive single molecule of DNA. The human genome contains about three billion bases and is small compared to some species genomes.

The 'code' is produced by triplets of bases (known as codons) standing for one of some twenty amino acids which are the building blocks of proteins. Proteins are long chains of amino acids. But first, the 'code' in DNA is translated into RNA which is built by the enzyme transcriptase. Just as DNA always cross–links in the double helix in a fixed way (A–T, G–C), so RNA is always transcribed as A–U, G–C, (uracil being RNA's equivalent of DNA's thymine). It is this RNA code that codes for amino acids when 'read' by the cell organelles, ribosomes.

Examples of this RNA amino acid code are UUU = phenylalanine; GCA = alanine; CAA = glutamine. For a complete list of these codons see the Wikipedia entry, *Genetic code* (https://en.wikipedia.org/wiki/Genetic_code).

8. Germinating the Tree of Life

One of the best books on this subject is The Ancestors Tale, by Richard Dawkins and Yan Wong (Dawkins & Wong, 2006). It takes us from humans back down the human branch of the tree of life, past the primates and mammals to the vertebrates and chordates, through the protostomes to the flatworms, the eukaryotes, archaea and bacteria. In this chapter we will come from the other direction.

Before I begin, there is another myth to dispel; the myth that modern human beings sit at the pinnacle of evolution; that somehow the purpose of evolution was to create humans and now it is finished. It is not. Every living species sits at the tip of their twig on the evolutionary tree and all have been evolving for exactly the same length of time.

No one species is more evolved than any other and certainly not any subspecies, race or regional variety. For example, some human populations are lactose tolerant; others are lactose intolerant. Both of those are perfectly normal conditions and neither is more primitive that the other because they are both adaptations to local conditions and niche opportunities.[1]

This book is about the human story in general and your story in particular. It could equally be the story of any other species or individual. It would only branch off at the point that that species' evolution branched off from the last common ancestor.

The next thing to consider is time. It seems incredible (there is that intuition thing again!) that from a simple self–replicating molecule, we have all this diversity of living things. It seems even more incredible when we consider that the distinct species alive today represent only about one percent of all the species that have ever lived.

It seems incredible because it is difficult for us to think in terms of deep time. We are more used to thinking in days, months and years so even things that happened in, say, 1066 seem to have happened a long time ago and the building of the first stone building, the pyramids at Giza, Egypt are so long ago that no–one is even sure how they did it now, and what they believed before they built temples like those at Karnak and why the pillars in the Karnak temples are shaped like lotus flowers is anyone's guess. Far, far too long ago! But on a geological time–scale, that was all very recent. That was practically yesterday!

There are probably few better ways to illustrate this that that used by Richard Dawkins in *Unweaving the Rainbow* (Dawkins R., 2006):

> *Fling your arms wide in an expansive gesture to span all of evolution from its origins at your left fingertip to today at your right fingertip. All the way across your midline to well past your right shoulder, life consisted of nothing but bacteria. Multi-celled invertebrate life flowers somewhere around your right elbow. The dinosaurs originate in the middle of your right palm and go extinct around your last finger joint. The whole story of Homo sapiens and our predecessor Homo erectus is contained in the thickness of one nail-clipping.*

> *As for recorded history; as for the Sumerians, the Babylonians, the Jewish patriarchs, the dynasties of the Pharaohs, the legions of Rome, the Christian Fathers, the Laws of the Medes and Persians which never change; as for Troy and the Greeks, Helen and Achilles and Agamemnon dead; as for Napoleon and Hitler, The Beatles and Bill Clinton, they and everyone that knew them are blown away in the dust from one light stroke of a nail file.*

It is probably worth reading that again, especially the bit about for how long there was nothing but bacteria, and how short has been the whole of human history and especially that which we call known history.

Very many of the fundamental internal cell processes that are essential to our lives and the lives of every living thing evolved in bacteria, including maybe the genetic code.

The third thing to understand is that when we talk about a process evolving, we are not talking about something that sprang into existence in a single instance in a single cell. We are usually talking about something that evolved slowly across the entire population, or at least a substantial part of it, with elements of it evolving independently of other elements, then later coming together to form something even more advantageous than the sum of its parts. Even if the probability of an advantageous change occurring are very small, we are talking about bacteria with a generation time often measured in tens of minutes in favourable situations. A hundred million bacteria replicating every hour very quickly produces the billion to one chance.

In addition, bacteria are known to swap genetic material in a simple form of 'sexual' reproduction, and they die and leave genetic material around to be taken in by other bacteria. The more successful bacteria will leave more of the good stuff that made them successful lying around in the environment. Then there is the role of viruses. Some viruses insert the DNA version of their RNA into their host's genome where it can become part of the host's DNA. So, if a virus incorporates the RNA version of an earlier host's DNA into their own genome, it can be transferred to a new host. The result is that a piece of DNA is now replicated in a new bacterium.

And all the while, generation after generation, natural selection is favouring those better at leaving descendants and quickly removing anything that makes reproduction less likely. Every generation is filtered through a sieve that allows more of the advantageous genes through and fewer of the less advantageous genes, while seriously deleterious genes are quickly eliminated. The effect is to concentrate the advantageous genes up an improbability gradient so what was once the billion to one chance becomes increasingly common in successive generations.

In addition to bacteria, biologists have identified another class of simple, single–celled organisms call archaea. These are similar to bacteria in having a single strand of DNA but they have a different cell membrane enclosing them. Because bacteria and archaea have just a single (circular) strand of DNA they are termed prokaryote cells. Scientists do not yet know for sure whether bacteria evolved from archaea or whether the two different prokaryote cells evolved independently from the first replicating molecules.

The reason this is important to you is that it is here that your oldest genes are being cooked up. Many of the basic processes that make yours and almost everything else's cells work are being created, not by magic although it is something almost magical, but by the simple process of evolution by natural selection.

Before we skip lightly over the next few billion years to where more complex cells, prokaryote cells, arise we need to go back one step. We need to look at how simple self–replicating molecules became bacteria and archaea.

The honest answer to this question, like the question of how the first simple self–replicating molecule arose and what it was, is that we do not yet know. We do not know if it was a single line of development or two or more that later got together. However, laboratory experiments have come up with a very plausible series of steps, as outlined in a New Scientist article by Nick Lane and Michael Le Page (Lane & Le Page, 2009). They assumed that the most likely location for it to have happened was in porous rocks in alkaline waters around geothermal vents and outlined ten steps:

1. Water filtering down into newly–formed rocks around geothermal vents reacted with minerals to produce an alkaline, hydrogen and sulphide rich fluid that welled up in the vents.

2. This fluid reacted with acidic sea water which was then rich in iron to form deposits of highly porous carbonate rock and a foam of iron–sulphur bubbles.

3. Hydrogen and carbon dioxide trapped in these bubbles reacted to make simple organic molecules such as methane, formates and acetates; reactions that would have been catalysed by iron–sulphur compounds.

4. The electrochemical gradient between the alkaline fluid in the pores and the acidic seawater would have provided energy to drive the spontaneous formation of acetyl phosphate and pyrophosphate. These behave like ATP (adenosine triphosphate)[2] which powers modern cells. This power supply would in turn power the formation of amino acids and nucleotides.

5. Currents produced by thermal gradients and diffusion within the porous carbonate rock would have concentrated the larger molecules creating the conditions for building RNA, DNA and proteins and creating the conditions for an evolutionary process where molecules

that could catalyse the formation of copies of themselves would quickly dominate and win the struggle for resources.

6. Fatty molecules[3] would have coated the surface of the pores in the rock, enclosing the self–replicating molecules in a primitive cell membrane.

7. Eventually, the formation of the protein catalyst, pyrophosphatase enabled the protocell to extract more energy from the acid–alkaline gradient. This enzyme is still found in some bacteria and archaea.

8. Some protocells would have started using ATP as their primary energy source, especially with the formation of the enzyme ATP synthase. This enzyme is common to all life today.

9. Protocells in locations where the electrochemical gradient was weak could have generated their own gradient by pumping protons across their membrane using the energy released by the reaction between hydrogen and carbon dioxide, so producing a sufficient gradient to power the formation of ATP.

10. The ability to generate their own chemical gradient freed these protocells from dependence on the pores in the rock, so they were now free to become free–living cells. This could have happened at least twice with slightly different cells, one type giving rise to bacteria; the other to archaea.

The above ten–step process is of course speculative and probably impossible to test and verify in a laboratory because the conditions around these geothermal vents deep below the ocean would be impossible to replicate in a laboratory, as would the time it might have taken. No–one is claiming it all happened in a day or two, or even weeks or years; not even the lifetime of a working scientist. It could have taken tens or hundreds of millions of years. No–one was in any hurry and there was no objective. Things happened when they happened.

The crucial point is that none of this is implausible; there is nothing in the laws of chemistry and physics which would prevent it. If chemistry **can** happen it will happen.

So now we know that the evolution of primitive bacteria–like and archaea–like cells was not only possible but entirely plausible. We do not know the precise details, but we know it happened because there are bacteria and archaea around today.

We have now gone from simple, self–replicating molecules to simple prokaryote cells. In a few billion years these will have evolved and refined and perfected their internal chemistry, all the while driven by a simple test of fitness to replicate themselves. Those that were better at it came to dominate those that were not so good, in their particular environment. In a different environment, others came to dominate.

Incidentally, did you notice how in all of that, nothing but chemistry and physics was happening? There was nothing magical called 'life' that needed to be added. So where did 'life' begin? Did it begin with the first replicators; with the first synthesis of energy storage molecules; with the first primitive membranes or the proton pump? Or was it when the first protocells broke free from the rocks and became independent, self–replicating organisms in every biological sense of the word? Nowhere, of course! 'Life' is chemistry. It doesn't require magic to explain it. And nor do you! You are much more wonderful than a conjuring trick.

We are now, on Richard Dawkins' outstretched arms analogy, having travelled all the way up our left arm and across our chests, heading past our right shoulder down towards our elbow. But before we got there, we went past a stage in the evolution of bacteria that was massively important to our story and the story of life on Earth. We went past the point where a bacterium learned how to use sunlight as more than just a source of heat. It learned to use sunlight to turn carbon dioxide and water into the simple sugar, glucose.

And in doing so it almost poisoned everything on Earth to death with the toxic by-product of this process, free oxygen.

Before this, Earth's oxygen had been bound up as oxides of other elements. Being highly reactive, in the hot conditions at Earth's formation, atmospheric oxygen would have all burned up to form oxides like carbon dioxide, silicates, sulphates, nitrates and especially hydrogen oxide – water.

A form of photosynthesis had probably evolved earlier, using chemicals like hydrogen or hydrogen sulphide as the 'reducing agent' (the chemical that donates electrons to the reaction). What these new bacteria did was find a way to use water – from whence the toxic oxygen came.

The evidence of this sudden presence of free oxygen is that the sea went 'rusty' due to the iron dissolved in it being oxidized. It then sank to form distinct iron oxide bands in the ancient ocean floors.

These bacteria were the cyanobacteria and they had evolved photosynthesis using an interesting enzyme that goes by the chemical name ribulose-1,5-bisphosphate carboxylase/oxygenase, mercifully shortened to the commonly used name, RuBisCo.

RuBisCo is probably the most abundant protein on Earth and there is a particularly good reason for its abundance. It is horribly inefficient. It appears in every cell of every green part of every plant, in the cell organelle known as the chloroplast. Here it is an essential part of the chemical process that takes a photon of light to start a chain reaction, the end point of which is a molecule of glucose. The chemical formula for the overall reaction is:

$$6CO_2 + 6H_2O \xrightarrow{\text{sunlight}} C_6H_{12}O_6 + 6O_2$$

In other words, six molecules of carbon dioxide plus six molecules of water, with added sunlight, makes one molecule of glucose and six molecules of oxygen.

The problem with RuBisCo is that not only is it very slow, typically only being able to complete between three and ten reactions a second, compared to thousands of reactions a second for some enzymes, but it makes lots of mistakes. One mistake is to produce chemicals like xylulose-1,5-bisphosphate which actually inhibits RuBisCo. Another is to mistake an oxygen atom for a nitrogen atom and incorporate that instead into the reaction. This causes the entire chain reaction to collapse wasting both time and energy. RuBisCo is very much the weakest link. Everything else is limited by RuBisCo's limitations.

The problem is that RuBisCo evolved in an oxygen–free atmosphere where these mistakes couldn't be made. Now there is lots of oxygen sloshing about, produced in part by RuBisCo, it slows everything down. In the early days, of course, when there was little free oxygen, being able to use water as the reducing agent, even very inefficiently, would have been much better than the alternative.

Now everything that uses photosynthesis as its energy source, and that means not just plants but everything the eats plants and everything that eats herbivores, depends on RuBisCo and there is nothing in evolution that can redesign it. Any deconstruction prior to reconstruction would make it even less efficient because evolution just doesn't have a reverse gear. So, we are stuck with RuBisCo and its inefficiencies.

RuBisCo is probably one of the best arguments against Intelligent Design, because it simply was not intelligently designed with any plan or foresight especially of the consequences of all that free and reactive oxygen. It is a notable example of the unthinking, unplanned nature of evolution where the only test is the utilitarian one – does it work at the moment and is it better that what preceded it?

As I said earlier, the cyanobacteria that first evolved photosynthesis using water as the electron donor almost killed everything with the oxygen they gave off. This was the first great mass extinction event.

As well see later, these early cyanobacteria are still with us, having later been incorporated into plant cells to give algae and eventually mosses, ferns and flowering plants. Green Earth owes its existence to these bacteria, as does every oxygen–breathing organism. You and I and all the life we see are descended from the few survivors of this mass extinction and the first example of serious environmental pollution by waste products.

And we survived because another group of bacteria learned to use oxygen to break down the very glucose these cyanobacteria had learned to make. These bacteria, the ancestors of the mitochondria in our cells evolved the ability to take a molecule of glucose, break it down in a controlled fashion and use the energy in it to build ATP from ADP (adenosine diphosphate) and phosphate. ATP is the main energy source for cell metabolism. As we saw in the ten–step process of possible abiogenesis, a form of this reaction probably evolved before there were even protocells. These ancestors learned to use toxic oxygen to their advantage.

Other bacteria were learning tricks like how to use pre–existing structures to make new structures such as the flagellum and cilia with their proton motors, the nearest thing to a wheel in nature, so they could swim around in their watery world.

Now let's look at what happened as we went past our right shoulder and were heading down to our right elbow.

Prokaryote species were now swarming in almost every imaginable niche on the planet, including extremely hot and extremely cold environments, as well as living in rocks deep in the ground. They had learned independently to make sugar from carbon dioxide and water and they had learned to use the free oxygen to burn that sugar to make an energy store to be used on demand later. They were motile with flagella and cilia and they were already in an arms race with viruses.

It would have been astonishing if some of them had not learned to live as parasites inside other prokaryote cells.

Parasites are extremely common in biology because the advantage to a parasite of living off the efforts of another, are obvious. The problem though (for the parasite) is that the host now has a vested interest in defending itself in any way it can, so an evolutionary arms race will ensue in which neither host nor parasite is the ultimate winner. A far more advantageous situation is a cooperative arrangement in which both host and parasite gain. This arrangement is known to biologists as symbiosis.

It was the evolutionary biologist, Lynn Margulis (Margulis, 1998), who in the 1960s began popularising an idea, first proposed as long ago as the 1910s by the Russian biologist and lichen expert, Konstantin Mereschkowski. He suggested that, as with lichens which are symbiotic colonies of fungi and bacteria, the complex cells might be symbiotic colonies – the endosymbiotic theory.

This idea, which was resisted at first, is now regarded as mainstream biology with emerging evidence from genetic studies that, for example, mitochondria are the descendants of once free–living bacteria that first parasitized other cells and then formed a symbiotic relationship with them. Imagine you're a cell making do with whatever the precursor to the Krebs' Cycle was for making ATP, and suddenly you find yourself infected with bacteria that can make ATP in abundance in return for some of your sugar and oxygen? What evolutionary advantage would there have been in fighting off the infection?

Similarly, you're a simple plant cell and along comes a cyanobacterium which takes up residence inside you and starts supplying you with lots of lovely sugar all for the 'rent' of some of your waste carbon dioxide and water. All you have to do is sit in the sun and get fed! Lovely jubbly!

It is now generally accepted that at least mitochondria in all eukaryote cells and chloroplast in plant cells are derived from bacterial parasites that later became symbionts and eventually merged with the original cells to become what we now think of as single entities.

The endosymbiotic theory explains how, millions of years later, these symbiotic organisms have become the organelles of eukaryote cells, including possibly the flagella and cilia of our motile cells such as sperm and the ciliated epithelia that lines our respiratory system and fallopian tubes. Most, if not quite all the DNA from these bacteria, has now been incorporated into a cell nucleus. The exception being the mitochondrial DNA which mitochondria still contain.

In maybe the most significant developments since abiogenesis, prokaryote colonies became eukaryote cells. In a very real sense of the term, eukaryote cells are still colonies of prokaryotes, and multicellular organisms are colonies of eukaryotes. The inescapable conclusion is that multicellular organisms are colonies of prokaryotes, bacteria and maybe archaea.

We still live very much in a bacterial world.

And the scene was now set for a radiation of a new type of organism – colonies of eukaryote cells – multicellular organisms.

End notes:—————————————————

[1] Lactose intolerance evolved in humans and other apes to wean babies off breast milk. Breast feeding acts as a natural contraceptive by causing the pituitary gland to release a hormone that inhibits ovulation. The benefit of this is that the baby gets the full attention of the mother for a longer time than if she became pregnant again soon after giving birth. But, unless this process ended at some point, there would be far fewer babies, so an inability to digest the sugar in breastmilk after about 18 months means the baby begins to reject breast milk, the mother begins to ovulate again. Evolution has resulted in a balance between the baby having more care and attention and more babies being born. In actual fact, rather than lactose intolerance, what happens is the digestive enzyme, lactase, is not made any more and the baby cannot digest lactose.

Lactose tolerance evolved when humans domesticated cattle and suddenly had lots of milk. Now, they could wean their babies off breast milk and onto cow's milk and, more importantly, they could continue to drink it into adulthood, so there was a clear evolutionary advantage in people with domestic cattle, to evolve lactase persistence. Women could have babies more frequently and the babies could continue to be adequately nourished, as could adults.

There is a very clear correlation between the incidence of lactase persistence and the incidence of domestic cattle and pastoral agriculture. There are signs that this has happened more than once and in different ways in different populations. There is a correlation between lactose intolerance and the occurrence of the tsetse fly for the simple reason that the tsetse fly harms domestic cattle which are derived from Eurasian species where there are no tsetse flies and in which resistance to the parasite it transmits has not yet evolved, as it has with native, African species.

[2] Adenosine triphosphate (ATP) is an adenosine molecule with three phosphate groups in a chain. Adding the third phosphate takes energy which is then available to power other internal cell reactions by breaking it down to adenosine diphosphate and phosphate. ATP is thus the cell's primary energy store. All cell metabolic processes reduce one or more molecules of ATP to ADP and phosphate. ATP is itself built in the mitochondria by breaking down

End notes – continued:

glucose in the presence of oxygen in the process known as cell respiration, also called the glycolytic pathway or the Krebs' Cycle, after the Nobel laureate, Hans Krebs who discovered it. Keeping cells supplied with oxygen for this process is the reason we have a respiratory and circulatory system.

[3] Fatty molecules or lipids have part of the molecule which mixes with water (lyophilic, solution–loving) and part which is repelled by water but mixes with oils (lyophobic, solution hating). They are capable of forming a double membrane with the lyophobic ends pointing inwards and the lyophilic ends pointing outward, held in a stable formation by electrostatic forces.

9. Onwards and Upwards

Here is where our story takes its first major branch off the rest of the evolutionary tree. Plants, and fungi now branch off and bacteria and archaea continue their own lines into their own futures. Our story is now that of the Animalia kingdom. Most of the spadework had been done laying down the foundations for our future by single cells iterating trillions of times round the reproductive cycle and filtering at each pass through the selection sieve.

Our ancestors were getting good at surviving! We owe them everything!

Multicellularity is thought to have evolved not just once or twice but at least forty–six times (Grosberg & Strathman, 2007) (Parfrey & Lahr, 2013) and not just in eukaryotes[1]. Even a few prokaryotes are also believed to have formed multicellular colonies.

Evidence of the first multicellular organism in the fossil record, at least the earliest so far found, was found in the Ediacaran Hills, Australia, from which the Ediacaran biota takes its name. This biota was the first flowering of multicellularity consisting of tube–like and frond–like, mostly sessile[2], organisms. These emerged soon after Earth came out of a period of global freezing for the second time. An earlier period had occurred just before the cyanobacteria evolved photosynthesis about 2.3 billion years ago.

The Ediacarans radiated and spread world–wide in the so–called Avalon explosion. They lasted from around 635 million years ago until about 542 million years ago, when they were quickly replaced by more mobile multicellular organisms in the so–called Cambrian explosion.

The Cambrian explosion was both a flowering of early multicellular organisms and a mass extinction, in that the Ediacarans quickly disappeared, presumed to have been eaten by the more mobile organisms. Having been almost entirely sessile, and having no mobile predators, the Ediacarans had not needed to evolve defences and would have been ready prey to animals with digestive systems capable of dealing with large organic molecules.

The Cambrian 'explosion' (almost as bad a misnomer as the Big Bang) is often waved around by creationists who like to present it as though it almost literally was an explosion in which lots of different creatures with lots of different body plans suddenly appeared, virtually overnight. It was nothing of the sort, of course. It actually lasted for 20–25 million year, starting 541 million years ago.

It was brought to popular attention by Stephen J Gould who used it to argue in his book, Wonderful Life (Gould, 1990), for his 'punctuated equilibrium' theory of evolution. According to this theory there are prolonged periods of stasis 'punctuated' by brief periods of rapid evolution.

Many people feel however that this was an attempt to explain an illusion caused by the apparent compression of time in the geological column, the relative rarity of fossilisation and the uneven rate of evolutionary change across a wide geographical range. Although the rate of evolution will be expected to vary due to varying rates of environmental change, the underlying mechanism is accumulated small changes over time, as the standard model predicts. At the local level, a more primitive form of an evolving species can be replaced very quickly by a more advanced form when two formerly isolated populations come into contact.

The reason it looks like a sudden event in the geological record is almost certainly because so many of the species were soft–bodied and only fossilised in very unusual circumstances. One such was the formation of the Burgess Shale in which most of the new species have been found. This appears to have been due to some highly unusual rock formations which resulted in periodic catastrophic flows of fine mud, either carrying the soft–bodied organisms with it or burying them where they were. These unusual conditions resulted in a unique form of fossilisation which managed to preserve some if not all, the soft body parts of the Cambrian biota. (The Burgess Shale Site 510 Million Years Ago)

It is hardly surprising therefore that this rare, maybe unique, sequence of events conspired to make it look like all the different phyla evolved very quickly and with no obvious ancestors. It is a very incomplete record. In 2013 an Australian team produced data showing that, although the Cambrian was a period of rapid evolution, "The fastest inferred rates are still consistent with evolution by natural selection and with data from living organisms." (Lee, Soubrier, & Edgecombe, 2013)

Be that as it may, the Cambrian was a period of what can be thought of as 'experimentation' with various basic body plans for a multicellular animal. At the end of this period there were just a few winning body plans, one of which was that of animals present in the Early Cambrian, the ancestor of the chordates and the ancestor of you and me.

Chordates are animals with a body plan that includes a notochord or stiff dorsal nerve cord. They are deuterostomes in that their anus develops before their mouth and they are also bilaterally symmetrical coelomates[3]. One group of chordates are the tunicates which includes animals such as sea squirts. These are placed in the chordate group because their juvenile forms have a notochord. This might give a clue to the origins of the chordates from a neotenous[4] tunicate. Two of the Cambrian chordates are vertebrates and regarded as early fish. Vertebrates are animals with backbones in which the notochord of the chordates is now a vertebral column in which a spinal cord runs the length of the body.

So, during the Cambrian, the basic design of your body had been formed and everything else was to be a variation on that basic plan. Your vertebrate ancestor was segmented; it had a backbone with nerves running in the spinal cord, a head with the main control centre, pharyngeal slits, a hollow body with a tube running from the mouth at one end and an anus at the others. And it had a tail extending beyond its anus.

It had a respiratory system for extracting oxygen out of the water it swam in and expelling the carbon dioxide it made as a by–product of its metabolism. It had a circulatory system for supplying oxygen and nutrients to its cells and getting rid of waste. And it reproduced by producing single–celled gametes which fused to make a new cell from which a new individual developed.

One problem that multicellular organisms which are more than simple colonies of otherwise identical cells have, is that they need to differentiate their cells into various specialised cells but with the same genetic code. Then they have to reset the entire process when two gametes fuse to make a new individual. It was probably solving this problem by the only method available – that of trial and error - that took evolution so long to come up with full multicellularity with the body arranged in specialised organs.

The still developing science of epigenetics[5] is making progress with understanding how this works and how one of each pair of chromosomes is effectively switched off, along with all the other genes that are not needed in a particular specialised cell. Once specialised, cells do not generally become generalised again but their inactivated genes are carried into all the descendants of those cells (Carey, 2012). Cell biologists know how genes are deactivated but not yet how the system is controlled. It is not yet known how the zygote that results from the fused gametes resets the cell so it has the full potential to become any cell in the developing embryo.

But whatever the controlling process is for the epigenetic system, it seems to have evolved very early on in the Animalia kingdom. We have what was established at least by the Early Cambrian. So our common ancestor emerged from the Cambrian Era with the basic vertebrate body plan, division of labour with specialised cells and organs and the potential for making every vertebrate alive today, including you.

Onwards and Upwards

End notes:———————————————————

[1] According to Wikipedia (Wikipedia), prokaryotes known to have formed colonies are *cyanobacteria, myxobacteria, actinomycetes, Magnetoglobus multicellularis* and *Methanosarcina*. The first evidence of multicellularity is by cyanobacteria–like organisms. Multicellularity has evolved in six eukaryote groups, animals, fungi, brown algae, red algae, green algae, and land plants. It evolved repeatedly for *Chloroplastida* (green algae and land plants), once or twice for animals, once for brown algae, three times in the fungi (*chytrids, ascomycetes* and *basidiomycetes*) (Niklas, 2013) and perhaps several times for slime moulds and red algae (Bonner, 1998).

[2] Sessile organisms are generally sedentary, lacking the means of locomotion.

[3] 'Bilaterally symmetrical coelomates' basically means they have a right and left side that are mirror images of one another, and they have a hollow body.

[4] Neoteny is the retention of juvenile characteristics into adulthood. The chordates may have evolved from the larval form of an early tunicate.

[5] Epigenetics is the study of a system that sits alongside the genes and deactivates unneeded genes in different specialised cells. A characteristic of multicellular organisms, especially animals, is that cells become highly differentiated and specialised, needing fewer genes to carry out their specialise functions. Control includes attaching methyl groups at key positions and binding stretches of chromosomes with clusters of four proteins called histones. Once deactivated, genes appear to remain deactivated for the life of the cell and are passed into all that cell's descendants so that, for example. liver cells always produce liver cells; muscle cells always produce muscle cells.

10. Walking with Dinosaurs

Your story has now branched again when, somewhere in the Early Cambrian, about 530 million years ago, the ancestors of bony fish began to diverge from the other chordates. They evolved a vertebral column from the notochord that surrounded the nerves running along their backs, and a cranium to house and protect a brain.

The first creatures that can be regarded as fish were jawless *Agnatha* (no jaw). These are all now extinct but have their modern counterpart in the lampreys and hagfish. Another fairly common little marine creature, the lancelet, formerly called *Amphioxus* but now renamed *Branchiostoma*, was thought to be ancestral to the vertebrates but is now believed to have diverged off the chordates at about the same time as the vertebrates. They are now regarded as a sub–phylum of the chordates, not ancestral to them. However, they are valuable to biology in that their genome shows examples of old genes that have been co–opted for new purposes.

But they are not our ancestors. Our ancestors had a well–defined head and a brain, and they had the beginnings of a bony skeleton. The reason this is significant is that bones not only give protection and to an extent shape, but they also allow directed and purposeful locomotion. It is no coincidence that the animals with true, jointed legs have a rigid skeleton of some sort. The arthropods, the group that include insects, mites and spiders, scorpions, millipedes, centipedes, woodlice, and crabs, prawns and lobsters, have an exoskeleton made of chitin with muscles inside hollow limbs. The vertebrates, the group that includes fish, amphibians, reptiles, dinosaurs, birds and mammals, have an internal skeleton, an endoskeleton with muscles on the outside of the bones. In the case of the cartilaginous fish (sharks, skates and rays) this is made of cartilage, and in the rest, bone.

The earliest fish came in two basic types, both jawless – an eel–like creature similar to the lampreys and hagfish, and a group of small, armoured fish, the placoderms. It was one of this latter group that first evolved a jaw, probably from one of its gill arches. It was the jawed fish that diverged into two major groups – the cartilaginous fish, or *Chondrichthyes*, and the bony fish, or

Osteichthyes. The bony fish in turn diversified quickly into the ray–finned fish, or *Actinopterygii*, and the lobe–finned fish, or *Sarcopterygii*. It was from these lobe–finned fish that our ancestors – the first terrestrial tetrapods – evolved. One group of lobe–finned fish are the coelacanths. This means we share a more recent common ancestor with them than either of us share with the sharks, skates and rays.

The basic quadruped limb plan was based on structures evolved in these lobe fins. The basic plan is a single long upper bone, jointed to two long lower bones then an arrangement of bones in the writs and ankle, then a five–digit hand or foot. This is modified according to need, especially in the number and length of digits, but the basic plan is still five. The lower 'leg' of a horse, for example, is actually a single digit. Even such highly modified limbs as the whale flipper and the wings of bats, birds and the extinct pterodactyls are modifications of this basic plan.

From that small, un–remarkable member of the Cambrian biota; those stiff little chordates that swam around amongst the trilobites and fearsome *Anomalocaris* with its huge, jaw–like appendages that were once thought to be a different species in their own right, big things were coming; in fact the biggest things to ever live – the blue whale. From that unpromising beginning were to evolve the fastest running animal, the fastest flying animal, birds that can soar on the wind and sing symphonies, three forms of powered flight, echolocation at least twice, a brain capable of doing calculus without realising it, and an ape that can go to the moon.

But first, it had to learn to walk on land and breathe air.

To walk on land there had to be survival advantage in doing so. It had to give access to more food, or sanctuary from predators. The probability is that what gave the first fish an advantage when they came out onto land was the same thing that gives modern mudskippers an advantage – lots of small arthropods and other terrestrial animal life to eat.

The earliest known land invertebrate was a species of millipede known as *Pneumodesmus newmani* that lived about 428 million years ago. It was found by an amateur palaeontologist and bus driver from Scotland named Mike Newman, hence its specific name. It is important to evolutionary biologists in that it has numerous holes in its cuticle that are probably spiracles. These are the tubes which lead into the bodies of terrestrial arthropods through which oxygen is taken and carbon dioxide is expelled. These spiracles are a strong clue that it lived on land and breathed air.

Before arthropods such as *P. newmani* could move onto land there needed to be some sort of ecosystem for them to move into. There would have been precious few advantages crawling about on bare rocks or sand, and soil needs plant matter, bacteria and fungi.

Evolution and ecology are really aspects of the same thing because ecology is about how species adapt and interrelate to form a complete system of interdependent organisms. Without evolution, none of the interdependent organism could fit into their particular niche and compete with other organisms for resources. So, evolution created ecosystems and ecosystems drive evolution. But in the early days of life on Earth, before life emerged from the oceans where it had first evolved and diversified, there was no ecosystem as such on land, so how did the first emerging life manage to eke out an existence and what reason did it have to go there in the first place?

In 2016, scientists believe they found evidence (Jinzhuang, et al., 2016) that an early plant, related to the club mosses, may have created its own ecosystem by interaction between its rhizomes and sediment from silt-laden floods. Rhizomes are underground stems by which many plants spread to form large colonies. Plants like this could have lived in shallow water prone to frequent drying so rather than the plants leaving the water, the water left the plants high and dry at times. It is but a small step then to spread out into 'dry' land prone to frequent flooding.

So, our intrepid millipede had some reason to move onto land in the form of plants to eat, as did others, followed no doubt by predators. Together with the early plants and maybe starting in estuarine tidal mudflats where there was an advantage in being able to survive dry periods, they created soils and ecosystems and niches for other creatures to emerge from the sea into or to evolve *in situ* from earlier colonists.

And one of these was our ancestor, an advanced lobe–finned fish that had evolved limbs with which to crawl around on the bottom of the sea where its body–weight would have been partially supported by water. The other thing it had evolved was the ability to breath out of water. This was maybe not as big a step as it might intuitively seem, and breathing air did not require the fish to live out of water.

The early fish, probably the armour–plated placoderms, had evolved the swim bladder as an aid to buoyancy from an outgrowth from the gut. As so often with evolution which has no hesitation to use a structure evolved for one purpose for something completely different, this swim bladder was probably co–opted to act as an oxygen store by inflating it with air gulped from the surface. In the early Devonian, oxygen levels were only about half of today's value, but a given volume of air contains much more oxygen than the same volume of water so anything that could access this atmospheric oxygen would have been at an advantage. It would have been especially advantageous for a lobe–finned fish that crawled about the bottom, foraging in the deoxygenated silt.

So, any fish that could haul itself out of water and 'breathe' atmospheric oxygen, even for a short time, could also eat the terrestrial invertebrates that had got there first, and those that could do it better and for longer and which could move faster would do better than the slower ones still tied to the water and unable to venture far from it.

Despite the creationist mantra that there are no transitional fossils, there are actually a large number of fossils showing the progress of the lobe–finned fish into terrestrial tetrapods. One of the best-known examples is *Tiktaalik*, the discovery of which represents a successful prediction of the Theory of Evolution. It was discovered by Edward Daeschler of the Academy of Natural Sciences in Philadelphia, Neil Shubin of the University of Chicago, and Farish Jenkins of Harvard University in Cambridge, Massachusetts. They had worked out that there should be a transitional species between the lobe–finned fish and the terrestrial tetrapods (amphibian, reptiles, birds and mammals) and that those fossils should be found in Devonian rocks.

Noticing that just such exposed rocks were to be found on Ellesmere Island, in Northern Canada, they assembled a team to go and look for them. Because of their remoteness and inaccessibility, only being accessible by plane, the rocks had never before been explored for vertebrate fossils.

They found several well–preserved specimens of exactly what the theory predicted should be there; *Tiktaalik*, given an Inuit name for a burbot – a freshwater relative of the cod – suggested by Inuit elders in Canada's Nunavut Territory, which includes Ellesmere Island.

The apparent gap in the fossil record between the lobe–finned fish and terrestrial tetrapods is still referred to as Romer's Gap, named after the palaeontologist Alfred Romer who first recognised it. It was substantially closed and shown to be more due to a lack of discovered fossils rather than an actual lack of them, by a team from Canada and the United Kingdom (Anderson, Smithson, Mansky, Mayer, & Clark, 2015).

The fossil record is never better than the latest discovery. A detailed examination of rock deposits at Horton Bluff, lying along the Avon River near Hantsport, Nova Scotia, Canada, showed abundant fragments of intermediate fossils showing both features of lobe–finned fish and the basic skeletal pattern of terrestrial tetrapod limbs. They differed in the number of digits, showing that the five–toed basic pattern of the general tetrapod limb had not yet evolved.

If you could ever get a creationist to describe what they would expect a transitional species midway between a lobe–finned fish and a salamander to look like – which is about as easy as drawing blood from a stone – they would be hard–pressed to describe something that was not a lot like *Tiktaalik* or the fossil remains found at Horton Bluff. Another gap in the story is closing fast.

So, our ancestors learned to live on land, to breathe air and to walk on four legs and we left our fish cousins to evolve on their own branch of the tree of life as we branched out on our own. The world of dry land where plants and invertebrates were already evolving and diversify lay before us, full of opportunity. First though, our ancestors were still tied to the water and wet places because their skins needed to be kept moist and for another particularly good reason – reproduction.

Fish and amphibians do not generally have internal fertilisation. The male doesn't need to penetrate the female in order to have his sperms fertilise her eggs because he can shed them onto her eggs in the water in which she lays them. The young develop in the water and, in the case of amphibians, go through a distinctly fishy larval stage or tadpole. So, the next trick our ancestors needed to learn was how to fertilise an egg and have it develop out of water.

Only one amphibian, one of the so–called reptiliomorphs (reptile–like) amphibians seems to have managed this trick so the entire clade of reptiles, birds and mammals have descended from this single species. These are collectively known as the amniotes, after the amniotic sac that surrounds the developing embryo, either in an egg or internally as with placental mammals.

The tetrapods that learned this trick evolved about 312 million years ago in the Carboniferous period. How exactly they did it is one of the challenges of evolutionary biology, made more difficult in this case because changes in the structure of eggs and of internal organs tend not to leave much in the way of fossil records. Nevertheless, we can make a few quite reasonable assumptions about what might have happened. There seems to have been two stages in the process, the first creating the possibility of the second.

It is distinctly advantageous to a male to increase the chances that his sperms are the ones that fertilise the female's eggs, so anything which enhances this likelihood would be favoured by natural selection because it would produce offspring carrying the genes for doing so. One way of doing this for a species that needs to return to the water to lay eggs would be to fertilise the eggs out of the water when they are still in the female. Depositing sperm into her oviduct would mean the eggs are fertilised before being laid (and thereby made available for the sperm of another male to fertilise them).

Now, assuming this method of fertilisation evolved – and if it did there would now be an evolutionary advantage in evolving more efficient genitalia such as a penis, mating rituals, female sex–selection, etc. – there were now opportunities for the female to do something other than to lay her eggs in water, mostly as food for other creatures. She could retain them for longer and lay them partially developed, so becoming active much more quickly after being laid. She could even retain them until fully developed juveniles and 'give birth' to them on land. She also needed only to produce a small number of eggs, not the hundreds or thousands that fish and amphibians need to lay in order for one or two to survive to adulthood.

There was also now the possibility for the second stage: to evolve an egg that could be laid on land. All it needed was a protective tough skin (shells came later) through which gasses like oxygen and carbon dioxide could diffuse; a water supply and a food supply for the growing embryo. Eggs already had the beginnings of these of course.

What the early amniotes evolved was a more complex system of membranes, some of which came from the early stages of embryonic development – the amnion, so that the embryo developed inside a sack or its own creation. One of the advantages of this was not only greater protection but a retaining membrane for a much larger yolk, meaning a much larger individual could emerge from it, and a reservoir for waste. It is this amnion that distinguishes all the amniotes from the anamniotes (amphibians and fish) and gives the clade its name. All the amniotes are regarded as a monophyletic group in that they are all thought to have evolved from a single common ancestor.

So, during the Carboniferous geological age, our ancestors became free from the need to return to water to breed and could move out into the land, exploiting the diversifying plant and invertebrate life and all the ecological niches that were opening up. We have left our cousins, the fish and amphibians behind to evolve in their own water–bound environment, and we are moving out into the world to see what it has to offer. We will soon find there is a lot of competition, some of it looking decidedly unfriendly. Earth was not always a fun place to be.

For a while, these newly–independent tetrapods had it good in the steaming swamps and tropical rainforests of the Carboniferous. The green plants pumped out oxygen into the atmosphere and the animals pumped out carbon dioxide that the plants used. The tetrapods quickly radiated into new niches and evolved into early reptiles, mammals, turtles, and crocodiles. The latter two even returned to the water but never returned to breathing with gills, nor did they abandon the tetrapod limb, although the marine turtles evolved flippers out of them. And they needed to return to land to lay their amniote eggs. They were aquatic land animals. Evolution has no reverse gear!

It was not to last. At the end of the Carboniferous, around 305 million years ago, Earth's climate went into a rapid, self–reinforcing decline for reasons which are not yet fully understood – the so–called Carboniferous rainforest collapse (CRC). Just as it was when the first creatures crawled out of the oceans, plant life is the foundation of ecosystems. The CRC devastated the plant life and with it, much of the animal life and our ancestors were threatened with another mass extinction.

The warm, steamy forests of tree–ferns that dominated the then single land–mass, Pangea, began to dry up and fragment, quickly leading to a loss of habitat for many creatures. Fewer animals meant less carbon dioxide in the atmosphere and, because of the loss of this greenhouse gas, a rapid cooling globally. The polar ice caps expanded locking up enough water to lower sea–levels dramatically, leading to a further loss of habitat for the amphibians left high and dry by the receding tides. This was a mass extinction which hit the amphibians hard but which was to benefit the surviving reptiles in the long run.

The reptiles, amongst them our ancestors, were much better equipped to cope with these new conditions, amphibians being so reliant on water both for breeding and to keep their skins moist. This gave the reptiles the upper hand and an opportunity to radiate into the new niches, while the amphibians became something of an evolutionary backwater – almost literally.

Fragmentation of the forests produced isolated populations in micro–ecosystems, giving a boost to the evolutionary radiation into new species. Reptiles became carnivores and herbivores as well as the insectivores they had mostly remained since the tetrapods crawled out of the oceans. Amongst them were the ancestors of mammals.

Fairly early on the amniotes split into two groups, the synapsids and the sauropsids. Mammals evolved from the synapsids; sauropsids evolved into dinosaurs. Dinosaurs came to dominate the land, evolving into very large herbivores such as diplodocus and triceratops and predators such as the tyrannosaurs, while the synapsids, mammal–like reptiles, remained small, barely exceeding the size of a rat. It is maybe worth reminding ourselves that, on Richard Dawkins' outstretched arm analogy we are now around the beginning of the fingers on your right hand.

With the dinosaurs and other reptiles dominating the daytime, the synapsids probably took refuge in the darkness of the night. This cowering existence was to stand them in good stead for what was to follow but first they had to adapt to the cold and make do with limited vision.

They probably evolved acute hearing to compensate for the lack of visual acuity in darkness; they probably evolved fur as an adaptation to the cold of night and they probably evolved warm bloodedness for much the same reason, and they probably lost colour vision as their retinas adapted to make better use of the shades of grey of the night–time. Evidence that very early mammals had hair, which was even showing signs of defensive evolution into spines and protective plates, was found in a fossil from Spain, the earliest known mammalian fossil, from 125 million years ago (Martin, et al., 2015).

These early mammals probably still laid eggs like their reptilian ancestors and much like the monotremes of today, the platypus and the echidna. Evolution of marsupials and placental mammals was to take another 70 million years.

What this adaptation to the darkness and cold of night probably gave to our ancestor was the ability to survive the cataclysm that finally finished off the large reptiles at the end of the Cretaceous when Earth was struck by a stray asteroid 66 million years ago. There is now little doubt that it was such a cosmic event that caused the mass extinction at what is known as the K–T boundary (K for Kretaceous (German) and T for Triassic) also called the K–Pg boundary (Paleogene).

Dinosaurs, which had been in decline for some time, quite suddenly disappear from the fossil record and there are signs of a sudden change in the weather for several years. Dust thrown into the atmosphere by a large asteroid explosively disintegrating on or near the surface would have plunged Earth into years of perpetual winter. This atmospheric dust has left a thin layer in marine sediment and terrestrial rocks which is high in iridium, a metal that is rare on Earth but common in asteroids.

We are now at the last joint of the middle finger of your outstretched right hand.

Because our ancestors survived that and lived to move into the niches previously the domain of dinosaurs, we are here.

Our mammalian ancestors weren't the only creatures to survive this mass extinction, of course. One group of dinosaurs, the theropods, which had evolved bipedal locomotion, so freeing up the forelimbs which could then be adapted for other functions than walking, had also evolved feathers from their reptilian scales. They had also evolved warm–bloodedness.

Like mammalian hair, theropod dinosaurs' feathers probably helped them survive the K–T cataclysm too. Being bipedal, their forelimbs were now free to be used for other things. They went on to evolve into birds. Sorry to spoil your romantic fantasy, creationists, but humans never walked with dinosaurs. It was to be another 64 million years or so before anything that could be called human appeared.

With a few exceptions like the Komodo dragon which had no competition on its island, the remaining reptiles and their descendants, the snakes, were to remain small. The mammals and birds were to go on to dominate the land and the skies. Some of them returned to the water but, like the turtles and crocodiles, always remained tied to the surface, at least for air. A few seals sometime give birth in shallow water near the shore but only two other amniotes have managed to remain permanently in water even to give birth, both mammals – the Cetacea (whales and dolphins) and the Sirenia (sea cows or manatees and the dugong). Our own ancestors will soon take to the trees.

11. Into the Trees and Beyond!

The first mammal that could be called a common ancestor of all the primates, tarsiers, tree shrews, lemurs, old and new world monkeys and apes, were rather late on the scene. In fact, by the time they arrived, most species of mammal to evolve had gone extinct.

Long before primates arrived, mammals had evolved the ability to do away with eggs altogether. They no longer just let the embryo develop partly with the nutrients in the egg before continuing to develop in a special pouch, like the marsupials do, but they feed and nourish the developing foetus in a specially adapted part of the reproductive tract, a uterus, via a special organ derived from the early stages of the embryo – the placenta. Before this could happen there was one major obstacle to overcome – the mother's immune system.

Probably from the beginning of life, living organisms have had to cope with other organisms trying to treat them as a food resource by living on and in them. In multicellular organisms, where the potential for parasitism is much greater, organisms would have evolved defences against them. At the same time, parasite would have been evolving ways to get around these defences in a classic evolutionary arms race – the existence of which surely argues strongly against any intelligent involvement in the process!

Humans, like other vertebrates, have a sophisticated immune system which detects 'foreign' bodies and mounts an attack on them by forming antibodies which attack and kill the invader. This is the reason transplant patients need to take special immunosuppressive drugs to prevent their transplant being rejected.

However, there is a class of viruses, which includes HIV, which have evolved the ability to suppress our immune systems, so preventing their rejection. These viruses are known as retroviruses because they have another neat trick to avoid detection. They insert the DNA transcript of their own viral RNA into our genome, making it look for all the world to your immune system like just another piece of our own DNA. Eventually, the organism manages to evolve a resistance to it, or the virus evolves, and the DNA Trojan horse is left behind;

sometimes a broken version of it, mostly doing nothing except being passed on to future generations and randomly mutating. It normally forms part of the junk DNA of which a great deal of an organism's genome consists due to copying errors which have duplicated stretches of DNA, for example – and of course these so–called endogenous retroviruses.

And one or more of these endogenous retroviruses might have been exapted to suppress the female's tendency to reject a developing embryo. Two scientific papers in 2012 showed how this may work (Haig, 2012) (Nancy, et al., 2012).

Incidentally, there can surely be few better pieces of evidence of the evolutionary process and how it produces nested hierarchies than these ancient endogenous retroviruses which can be arranged in cladistic hierarchies that exactly map onto the other genetic, morphological and physiological evidence for evolutionary relationships.

As with the way the K–T boundary disaster ended up presenting mammals and birds with a gift of vacant niches to exploit, this ability of evolution to make something good come out of something bad might seem almost mystical, even magical, until one realises how evolution always favours the good stuff; in other words the things that allow more descendants. The bad stuff quickly gets eliminated and consigned to the evolutionary trash can.

It is also an example of our tendency to assume that whatever allows us to be here now must have been for the long–term 'good'. Had our ancestors not survived the K–T disaster, then something else could have been here thinking what a wonderful thing it was that those creepy crawling hairy little rat–like things went too. If we had never evolved placental gestation, we might have been singing the praises of whatever our reproductive system would have been without the right endogenous retroviruses. There is no good or bad in evolution. There is just what is. It is the same mistake I discussed in Chapter 2 with the hands of cards. If we assume the intention was to evolve us, then these things become impossible to believe just happened by chance. Of course, as with the hands of cards, what we are looking at is how things turned out, not how they were meant to be.

The placental mammals now had a couple of possible reproductive strategies – they could either have a lot of small babies born relatively immature and nurture them outside the womb, or they could have a small number and let them develop for longer before giving birth to them. The former strategy allows frequent litters but carries a higher risk of the babies not surviving; the latter mean fewer babies but an increased likelihood of them reaching maturity.

Our early primate ancestors seem to have adopted the latter strategy.

The first known primate–like mammal is a 65 million year–old fossil from North America known as *Plesiadapis*. Similar 'basal' primates have been found in China and Eurasia.

The jury is still out on the question of where exactly our ancestor that went on to be the last common ancestor of the African apes and the other African primates diverged – somewhere in Eurasia or in Africa. I will cover that a little later on but first I will look at how we are the product of primate evolution from well before that divergence.

One of the consequences of having adapted to a nocturnal life–style when the dinosaurs dominated the day–time was the loss of colour vision. It made evolutionary sense then to have good black and white vision at the expense of colour vision, so our early mammalian ancestor lost two of the four cones in their retinas. So, our primate ancestors lived in a black and white world, but we now have colour vision again.

I'll go into this story in some detail because it nicely illustrates how evolution works, often as cooperative co–evolution between two unrelated species because it is mutually beneficial to cooperate. It is also part of your story and explains why you can enjoy a beautiful planet in glorious colour. This account is based on a blogpost I wrote after being inspired by a walk in autumn sunshine in Oxfordshire, England in 2013 (Rubicondior, 2013).

While the mammals had been revelling in their new–found freedom to try out a diurnal life–style a couple of parallel evolutions had been going on. Plants and insects, especially a group of *Hymenoptera* – bees, wasps and ants – had evolved flowers and pollinating insects which lived on nectar and transported pollen about, efficiently targeting receptive flowers of the right species at the right time. The co–evolution of flowering plants and pollinating insects is a fascinating story but alas not part of ours, at least not directly.

Meanwhile birds, with beaks but no teeth had evolved from the surviving feathered theropod dinosaurs.

One problem all plants face, especially the flowering plants freed by evolution from the restriction of needing more or less permanently damp conditions and able to grow in far more diverse habitats, is how to disperse their seeds. The last thing a newly-germinated plant needs is to find itself growing right next to mum, who will have dominated the local environment. Each plant of the same species is a rival for territory with every other one, and not always plants of the same species. There is no sentimentality involved; it matters not one jot to a plant that it is competing for survival with its offspring or sibling.

It is in an evolving plant's 'interest' to evolve dispersal mechanisms which scatter its seeds as widely as possible because that gives a better chance of producing at least one successful offspring during its lifetime (and that is all that is required for a stable population) than by simply dropping the seeds on the ground around it. So, many flowering plants have evolved fruit, which botanically is the apparatus for seed dispersal.

A walk in the country in the autumn almost anywhere in the world, especially where the seasons are fairly well demarcated, will reveal a mass of different fruits and berries, most of them fairly small (not all, but most). An English country hedgerow in autumn will have an abundance of hawthorn berries, rose hips of the *Rosa canina*, sloes, blackberries, elder berries and a dozen other small berries with the emphasis on the small. This is because they have a dispersal strategy which involved getting themselves eaten by birds.

In the case of the hawthorn, the strategy is to get eaten by a bird which digests the red-skinned fruit around the seed but leaves the hard seed unscathed to be dropped some distance away, complete with a little fertiliser to get it started. So, the fruits are small enough to be swallowed whole, so the seed isn't damaged, and attractive enough to be seen. A red haw is a ripe haw, and a ripe haw contains sugar and soft pulp, so birds like the common blackbird have evolved to recognise a ready meal sitting there waiting to be eaten. Most fruit turns colour very quickly when ripe, often in a matter of a day or two.

This dispersal process was possible because birds see in colour, and the colour the species the hawthorn needed to attract recognised red as the signal for ripe fruit. Additionally, for an animal with colour vision, red stands out, but for most mammals which see in shades of grey, the colour red is indistinguishable from green. Plants also evolved to hide their seeds from mammals with their crushing teeth.

Here then, as with flowering plants and pollinating insects, we have another mutually beneficial cooperation between distinct species. Plants get their seeds dispersed and birds get fed. This method could not have evolved before there were birds and very many bird species could not have evolved without fruit-bearing flowering plants; again, an example of an opened niche being exploited, then evolution continuing in a new direction that this opened up.

This method is hugely wasteful of course, and hawthorn trees must invest vast amounts of energy into making a huge surplus of seeds to ensure a small number succeed, but that is all it does, hence the vast abundance of haws; far in excess of what would be needed if they all produced another hawthorn tree. So, wasting haws is in the interests of bird genes too.

So, we now know then that hawthorns and dog roses must have evolved after birds, which first appear in the fossil record around 150 million years ago. We also know that whatever is driving this process is neither intelligent nor moral because such waste would be both stupid and immoral when viewed as the product of an intelligent designer.

But why the emphasis on birds as the favoured method of dispersion? Well, apart from the ability of birds to cover much larger distances in the course of a day than mammals normally do, birds see in colour and so can respond to the signal that the fruit is ready for its seeds to be dispersed. Additionally, they swallow berries whole, so don't crush the seeds inside.

And that leads to the ultimate point: how come, if mammals do not see in colour, humans do?

The reason for that is that of all the mammals, only a few, and almost all of them primates, see the world in other than shades of grey, and the reason for that is that the world of trees in which the primates evolved was one of ripening, high-sugar fruit because birds and flowering plants had been busy creating it for their own mutual benefit for several million years.

Our ancestors found an open niche which they could exploit through the process of evolution by natural selection to either re-activate colour vision which they had earlier lost as wasteful in a nocturnal animal or to evolve it anew. Birds and flowering plants had created an environment in which it gave our simian ancestors an advantage; to put it bluntly, we parasitised the system of avian plant seed dispersal, by treating both the fruit and the seeds as food. Not all simians have colour vision of course because it serves no purpose in nocturnal species, so the fact that we have colour vision strongly suggests we evolved from a diurnal species.

Of course, evolution did not stop at that point and several mechanisms have evolved since by which plants exploit fruit-eating mammals, hence the abundance of tomatoes close to raw sewerage outlets. Many other plants have evolved other strategies since mammals evolved, like sticking their seeds to mammalian fur for example.

So, turning that round we can predict that the first monkeys with colour vision did not evolve until after birds and flowering plants had jointly evolved fruit. It is not rocket science to predict that it was also after the evolution of trees, of course.

How does this prediction measure up to reality? The evidence is not straightforward with molecular clock estimates at about 85 million years ago, fossil evidence at about 55 million and other estimates at about 65 million years since the first proto-simians evolved. However, these are all well within the timescale predicted by our theory.

So, the reason we have colour vision is because birds do and because flowering plants exploited that ability to get their seeds dispersed. Then our ancestors exploited that system to evolve eating fruit.

Becoming fructivores was to have another consequence – we need Vitamin C in our diets because, unlike almost all mammals not on our branch of the evolutionary tree of life, we cannot manufacture it. Most mammals have a four–stage process for manufacturing this essential vitamin if they do not take in enough in their diet. We have the first three of the four stages which work perfectly well but the enzyme needed for the fourth stage is broken by a simple mutation in the DNA that codes for it.

In most mammals the setup is simple: if they have too much Vitamin C in their diet their kidneys actively excrete it; if they do not have enough, they manufacture it in their liver. What happened at some time in our evolutionary history is that a fruit–eating ancestor was probably getting too much Vitamin C from their fruit diet so they had to excrete it and the manufacturing process broke, but this went unnoticed because dietary Vitamin C was never a problem. It might even have been an advantage in that the manufacturing process didn't need to be turned off.

And it is still not normally a problem for us and all the other descendants of the fruit–eating ancestor, because we normally still get enough in our diets – unless we over–boil our green vegetable, especially in copper pots which catalyses the oxidation of Vitamin C, or we go on long sea journeys or artic explorations and rely on a diet of dried and preserved food, lacking in fresh fruit and vegetables.

If having the first three stages of manufacture in the liver still active and fully functional but breaking the fourth stage, and then not reversing that error when needed, is an example of intelligent design, then someone is using a private definition of intelligence – one that includes blind stupidity and lack of planning.

So, back to our early primate ancestors.

The earliest primate–like mammals lived in what are now North America, Europe and Asia, which were then joined, between 65 and 55 million years ago. These small, tree shrew–like mammals had a generalised body plan showing few signs of specialisation. They still had five digits on each foot and were probably insectivores that could climb trees.

By about 55 million years ago there were some 60 different genera of primates, mostly from two families, the *Adapidae* which were similar to modern lemurs and lorises, and the *Omomyidae,* more like galagoes and tarsiers. This period, the Eocene, was something of a flowering of the early primates with about four times the diversity of today. Somewhere amongst the *Omomyidae* were our ancestors.

During the Eocene, some major evolutionary changes were occurring in the primate skull, indicating a change of life–style. Eyes were getting larger and moving more to the front of the skull, indicating stereoscopic vision, and the *foramen magnum*, the large hole in the skull that the spinal cord passes through was moving from the rear of the skull to underneath it.

The significance of the change in the position of the *foramen magnum* is that, in an animal which holds its head and particularly its face and eye horizontally in line with a horizontal backbone, the *foramen magnum* is at the rear or the skull. In mammals that are more upright with a face rotated down to still look forward, the *foramen magnum* is more underneath, so the head balances on top of an upright spine. Most of the monkeys, and especially the apes, tend to hold their body upright and look forward from this upright position. Together with grasping hands and feet, these features suggest our ancestors were climbing about in rocks and trees and were feeding hand to mouth.

At some point around 40 million years ago, the simians (monkeys) split into the Old World monkeys (the *Catarrhini* – flat–nosed) now found in Africa and Asia, and the New World monkeys (the *Platyrrhini* – down–nosed) now found in South and Central America.

There are two ways in which the *Platyrrhini* could have ended up in South America – either they were present in North America and migrated down the Isthmus of Panama when North and South America joined, or they somehow got across the South Atlantic. South America and Africa were much closer then, the Atlantic drift not having progressed to today's extent, but that still leaves either rafting on floating vegetation, maybe island–hopping when sea levels were lower, or a land bridge. There is no fossil evidence to suggest a North American origin, so a combination of rafting and island hopping is still the favoured explanation.

The lack of diversity in the New World monkeys – there are only five families in one super family, suggests the invasion of South America was a fairly rare occurrence, achieved by just the one progenitor of the supper family, the *Ceboidea*.

But our story continues now with the Old World, flat–nosed, monkeys, and in particular those that migrated down from Eurasia into Africa where the rest of our story until just a few hundred thousand, maybe a million or so, years ago was to be written. From now on, until the archaic humans migrated out again, we were an African species.

In 2013, a paper in *Nature* (Stephens, et al., 2013) reported the discovery in Tanzania of both the earliest known 'ape' and the earliest known Old World monkey both found in the same precisely dated stratum from 25.2 million years ago. The discovery was in the Rukwa Rift, a segment of the western branch of the East African Rift. This shows that the divergence was soon after the Old-World monkeys arrived in Africa or even before.

At some point, about 12–18 million years ago according to the molecular evidence, gibbons (*Hylobatidae*) split off from the other apes and ended up in South East Asia, as did the orangutan (*Ponginae*) which split off about 12 million years ago. There is no known fossil record tracing the split of the gibbons and their migration to South East Asia and there is only tenuous evidence of fossils from India and Turkey which could be early *Ponginae,* so it is possible that the South East Asian apes split from an even more remote common ancestor that was widespread across Eurasia, one descendant of which migrated into Africa.

The next major split in the African apes was about 7 million years ago when the gorillas split from the other hominids. These then split into the chimpanzees and the australopithecines about 4 million years ago. The chimpanzees quickly split into the chimpanzee (*Pan troglodytes*) and the bonobo (*Pan paniscus*) while the australopithecines diverged into several species, one of which became the *Homo* branch of the evolutionary tree.

The split between the chimpanzees and the hominids was probably caused by climate change which broke up the African forests in East and South Africa leaving our ancestors in isolated trees and grassland. This is what probably drove the evolution of a fully bipedal gait and the 'human' foot as a means of quickly moving from one patch of trees or rock outcrop to another.

While this book was in preparation a paper published in the online journal PLOS ONE cast doubt on some aspects of this assumed split between the hominins and the Great Apes occurring in Africa at all. It suggests, on the rather tenuous evidence of a single jaw with a tooth with a typically hominid fused root that the split could have occurred in Europe. The jaw of

Graecopithecus freybergi was found in Greece and has been reliably dated to between 7.11 million and 7.37 million years ago. (Böhme, et al., 2017)

This paper is in contrast to the findings of a French team from the University of Poitier, presented to the 2013 Annual Meeting of the American Association of Physical Anthropologists (Bienvenu, et al., 2013). This presentation reported that a skull from the Republic of Chad, *Sahelanthropus tchadensis*, (Sahel ape from Chad) dated to 7 million years ago, had a cranial capacity of a chimpanzee (panin) but a tilted brainstem and other hominid characteristics suggesting at least the beginnings of bipedalism. The skull is exactly what would be expected of a transitional species between panini and hominids.

The attractiveness of Chad, at the western end of the 'Sahel Corridor' as the site of this possible divergence is considerable. The Sahel Corridor runs south of the Sahara and has been a migration route for many species and human groups, across Africa and so forms a possible route for early hominids from the forests of West Africa where the other African apes are found, across to East and South Africa where the Australopithecines evolved into archaic and modern humans.

The 'human' foot at the end of a leg which is barely distinguishable from a *Homo sapiens* leg was certainly present in pre–human australopithecines, as can be seen by the famous Laetoli footprints from South Africa. These were discovered by Mary Leakey in 1976 and were made by about 3.7 million years ago, probably by *Australopithecus afarensis* or possibly by *Ardipithecus ramidus* or a close relative. *Ardipithecus* was an early australopithecine.

The footprints were made in fresh volcanic ash which was either wet at the time the prints were made or became wetted by rain soon afterwards, and then dried to set solid, preserving the footprints in sufficient detail for a detailed analysis to show the maker walked in an identical fashion to modern humans and therefore almost certainly had lower limbs very similar if not identical to modern humans.

Another discovery at Laetoli reported in eLife in 2016 (Masao, et al., 2016), showed many more similar footprints all going in the same direction. They also showed considerable variation in size, suggesting marked sexual dimorphism; males being considerably larger than females.

The evolution of the bipedal lower limb involved a skeletal adaptation in addition to the bones of the ankle, foot and leg bones of course. It required a rotation of the face to bring it round 90 degrees to the spine, not in line as with most mammals. We have already seen how this process began with the simians as seen in the movement of the *foramen magnum* but there was another change needed at the base of the vertebral column – in the pelvis.

The spinal column of the upright hominids needs to bend in the lumbar region to bring it upright too and this meant a change in the pelvis which was to have profound consequences for you and me later on. It may be the reason we are able to learn, communicate effectively and acquire culture, for example.

The problem was that the change in the pelvis caused the birth canal to bend too, so hominid babies are pushed through a curved birth canal rather than the straight, horizontal one of most mammals. This means that the head has to remain relatively small even when the baby is born relatively well developed as it is in the other apes. With humans, especially when our brains began to enlarge, this would have resulted in a rising perinatal mortality rate for both mother and child. Not good from an evolutionary perspective – unless the baby is born relatively under–developed.

But this makes it dependent on its parents (and maybe siblings) for much longer. It also means it can now grow a larger head than the mother could have given birth to. Result? A larger head in the long run and more prolonged period of nurturing as an infant. Evolution had chanced upon the recipe for producing clever adults, able to learn and remember skills from their parents and, more importantly, to pass those skills on to the next generation. Hominid learning might have become accumulative because we walked upright and had a curved pelvis.

Natural selection wins again! And you can read and wonder about it as a result.

But there was a long way to go yet, and probably many dangers. We had to cope with being fairly low down in the food chain of many predatory species – big cats, wild dogs, crocodiles, snakes; maybe even other species of hominid. We were relatively weak, relatively slow, relatively small and probably quite tasty. We were only ever going to survive if we stuck together, worked together and watched each other's backs. We had to learn to cooperate which meant we had to learn and obey the group's rules. I'll come to the evolution of social ethics and morality soon. First, we need to talk about the australopithecines – and whether they should be considered even a different genus to *Homo*.

One thing is certain though. Not only is there emphatically **not** an absence of 'transitional' species between apes and humans as creationists would have you believe; if anything, there are too many of them because what we have is not an absence of evidence but so much that it confuses rather than clarifies the picture.

Only a few years ago it seemed fairly straight forward – *Homo* evolved in the Afar region of Ethiopia from an australopithecine very similar to, if not identical to, *Australopithecus afarensis*. From the fossil remains of 'Lucy' it was clear that *Au. afarensis* was still partially arboreal but spent a good deal of its time on the ground. Scientists have even concluded from the fractures found in 'Lucy' that she probably died by falling out of a tree.

Now the picture has become more complicated. Not only are there a couple of other candidate species for the immediate ancestors of *Homo* but they come from South Africa, not Ethiopia or East Africa at all. *Au. sediba,* for example, is a South African australopithecine with a decidedly human lower body and the upper body and cranium more like a chimpanzee in size, except for the hands, which look human.

Then there is the recently–discovered *H. naledi*, so far found only in a single cave in South Africa, the Rising Star Cave system and again having a mosaic of human and chimpanzee features, including a small brain. Recent multiple dating methods, by independent teams, show that *H. naledi* lived in South Africa until some 300,000 years ago, when modern humans, *H. sapiens,* were certainly present and may well have been in South Africa.

In other words, an archaic hominid was probably living at the same time as modern humans and maybe even alongside them. It is unlikely to have been our ancestor, but the mystery is why it survived without evolving a large brain, if the evolutionary pressures that caused humans to evolve a large brain were present in South Africa. The 'Di*naledi*' chamber is so difficult to reach it is unlikely that the individuals found in it made their own way there voluntarily, suggesting a possible ritual interment there. Did these hominids with a brain little larger than that of a chimpanzee have some concept of an afterlife?

Despite these South African finds, the general consensus is still that the birthplace of the *Homo* genus was Ethiopia and that the major developments in human evolution were centred in East Africa, around the Rift Valley and the most likely ancestor was *Au. afarensis* or 'Lucy'. There are other contenders for this honour – *Au. garhi* and *Au. africanus* can also stake a claim.

Lucy was discovered in the Hadar region of Ethiopia in November 1974 by Tom Grey and Donald Johanson. They gave it the nickname 'Lucy' after the Beatles' song, *Lucy in the Sky with Diamonds*, which was popular at the time. Although the most complete skeleton, Lucy was not the first remains of *Au. afarensis* to be found; a knee having been found by Donald Johanson in the Afar Depression in Ethiopia, and a palate and upper teeth in October 1974. Subsequently, in 1975, the remains of thirteen individuals who all seem to have died at the same time, were found at Hadar. In 1992 a complete skull was found and in 2011 a foot showing a distinctive hominid arch was found at the Hadar site.

From these remains, it is clear that *Au. afarensis* is as good an intermediate between the australopithecines and *Homo* as you could wish for. It is clear now that, having split from the chimpanzees, the early hominids diversified in South and East Africa into a number of species of the *Ardipithecus* and *Australopithecus* species and that from one of these, the common ancestor of all the *Homo* species evolved, and most probably in East Africa where most of the archaic *Homo* species have been found, even if they have been found elsewhere.

What is maybe not so clear is why palaeontologists place a species like *Au. afarensis* in a different genus to *Homo*, when they have far more similarities than differences. Why aren't they all *Homo*, or why aren't all *Homo* species, including *H. sapiens* placed in the *Australopithecus* genus? This problem comes from the way taxonomists try to use the taxonomic structures that were first devised to classify living things, where the distinctions were more obvious. When it comes to classifying what are in reality, snapshots in the evolutionary history of a species, it becomes far more difficult and, in the end, arbitrary.

There never was a point in time where, had you been there, you could have said that *Au. afarensis*, for example, had become *Homo habilis* or whatever the first of the *Homo* genus was (another contender is *H. rudolfensis*). Evolution is accumulated small changes over time across the whole of the population. Just as the colours of a rainbow merge one into the other across the spectrum, so, with a few rare exceptions, species merge one into the other across their evolutionary history. Never would one individual be a different species to its parents and never would it be unable to interbreed with members of its parents' generation. A species is not one individual but a population which is genetically isolated and there is no way to tell if any one individual is going to be the progenitor of such a future population. There never would have been a last *Au. afarensis* or a first *H. habilis* so any line drawn in time will always be arbitrary.

Taxonomists and palaeontologists are not using the term 'species' in the same way that biologists do to classify living things. What they mean is that this find looks very much like some we have discovered already and significantly different to others, so we place it in this taxon. If it is not obviously one or the other, then it is given a new name, so everyone knows what they are referring to.

One consequence of this difficulty with taxonomy is that so many apparent species can cloud the issue rather than shed light on it. For example, some authorities separate *H. rudolfensis* and *H. habilis* into different species while others regard them as the same species. It all depends on what range of physical characteristics was normal across an interbreeding population and indeed whether they interbred. Given the evidence that interbreeding between different related *Homo* species was relatively common it is difficult to decide where to draw the line, if at all.

As we will see, even early modern human beings interbred with at least two related species and maybe with an archaic species in parts of its range. For much of our history, both *Homo* and *Australopithecus* probably acted more like a ring species, sometimes breeding when they over–lapped but otherwise acting like perfectly respectable species.

There is evidence that, as we split from the chimpanzee, there was a period of some two million years when interbreeding between the two diverging species occasionally took place when they came into contact (Patterson, Richter, Gnerre, Lander, & Reich, 2006). Divergence was not a sudden process but a gradual one taking two million years! There is also recent evidence of interbreeding between the two chimpanzees as they diverged. (de Manuel & et al., 2016). This is exactly what we would expect of course because no–one tells the diverging species that they are now distinct species and so to conform to human cataloguing systems, they should be remaining genetically isolated.

Another consequence of the difficulty with taxonomy in an evolving taxon is that as the point of emergence of a new genus from its immediate ancestor approaches, it becomes more difficult to classify them. For example, although *H. habilis* had a larger brain and smaller teeth than the australopithecines (normally regarded as the distinguishing characteristics) they in fact retained some of the arboreal characteristics of their immediate ancestors. Some authorities argue that this is sufficient to classify them as *Au. habilis*, not *H. habilis*. Of course, that made not one iota of difference to *H/Au. habilis* nor does it change our evolutionary history other than in an esoteric, academic way.

When I was about thirteen and living in the small North Oxfordshire hamlet I had been born and brought up in, I had been befriended by Professor Sir Wilfred Le Gros Clarke and his wife, who had had a house built in the village. Lady Freda Le Gros Clarke was a friend of my mother's and they had taken an interest in me because of my love of nature and reputation even then as something of a naturalist and fossil hunter. I used to take their grandsons fossil hunting in likely sites around the hamlet. I had been given free use of their extensive library; I also delivered their Sunday newspapers.

Sir Wilfrid had been one of the team that had proved the Piltdown Man to be a forgery and was a former president of the Royal society. He was considered the foremost paleoanthropologist of his day.

One Sunday morning I delivered the newspapers as usual and Lady Le Gros Clarke invited me in to meet a guest who had something that I might like to see. There was a golden house rule that shoes were left at the door, and I was wearing wellington boots, which gave me especially evil–smelling feet. By this time, I had walked round just about every house in the hamlet delivering newspapers and my feet were simmering nicely. So, being too embarrassed to take my wellingtons off, I made my excuses and declined the invitation; only to learn that the house guest who had something interesting to show me was Dr Louis Leakey, who had brought his famous 'Zinjantropus' (later classified as *H. habilis*) skull that his wife Mary had found in Olduvai Gorge, Kenya, to show to Sir Wilfred.

It was then believed to be the oldest ever hominid remains ever found and had been widely reported in the news at the time. I would have been one of the first people outside Africa to see it. *C'est la vie!*

Around 1.8 million years ago, *H. habilis* had evolved into *H. erectus*, possibly via *H. ergaster* which some authorities regard as a subspecies, *H. erectus ergaster*. *H. erectus* was probably the first human species to leave Africa. The combination of a larger brain, smaller teeth and signs of a shortened intestinal tract are strong indicators that *H. erectus* had learned to use fire for cooking and was probably living on a diet high in fats and protein. Growing a large brain imposes a very high demand for fuel which probably could not be provided by raw vegetables or even raw meat. Cooking food first makes it more digestible and so releases much more nutrient than does consuming it raw.

So, what was driving this 'demand' for a large brain? Or to put it another way, what environmental pressures were favouring a large brain even at the cost of needing high–quality food?

There are possibly two interrelated things that were going on in early humans as an enlarging brain itself created the potential for an even larger one to get better at what we were beginning to do. In addition to using tools, we were learning to understand narrative and use it to make predictions that were proving extremely useful. For example, humans are probably the only species capable of 'reading' the information in an animal track and to do this we need to be good at pattern recognition to see an animal track in the general background 'noise' of other marks in the environment.

Did this come from the need to recognise faces, or did our ability (even over–sensitivity) to recognise faces come from our ability to recognise patterns? We are certainly good at recognising a human face. We see human faces and human figures all over the place – on toast; in a dropped ice–cream on a pavement; in lichen on a wall… Jesus and Mary seem to pop up everywhere. No doubt Muhammad would too if it wasn't a profanity to depict him.

With practice and maybe training, we can look at an animal track and read some especially useful information in it, especially if we want to catch dinner and not be caught for dinner by something else. Those leopard tracks lead up to those rocks – do not go there. That gazelle is walking with a limp and went over there – could be an easy catch! A hyena was prowling round the camp during the night – best set guards or keep fires burning tonight.

We were also working together in social groups; probably the only way to survive in the hostile and threatening environment in East and South Africa. A paper published in *Nature Communications in 2013* explained how social groups will develop ethical behaviour over time as cooperative strategies predominate over selfish ones (Adami & Hintze, 2013).

Here we were, a puny ape, walking upright and able to throw sticks and stones, maybe even sticks with stone points tied to them as simple spears. We were surrounded by predators like lions, hyenas, cheetahs, leopards, crocodiles, pythons and venomous snakes and out there in the grasslands was food a-plenty on the hoof, if only we were big enough to catch and kill it and not get caught and killed by something else out looking for dinner. If not, we would have to depend on foraging for insects, lizards, roots and berries in season; more gatherers than hunters. We were never going to succeed as a species unless we worked together as a team, and we were never going to work together as a team unless we all understood the first rule of teamwork - trust.

Trust means you can depend on other team members to play their part, and they can depend on you. Betrayal of trust would have meant exclusion and almost certain death. To be in the team you had to do your bit. That might have meant simply being a look-out for predators looking to catch their dinner. It might have meant being a beater who drove the gazelle towards the spear-throwers. It might even have meant being the one who made the spear tips and tied them on the sticks. Or it could have been the ones who stayed behind and looked after the children.

Trust meant you got your share of the kill because you had played your part. Trust meant that the handful of people who actually made the kill didn't make off with it and keep it all for themselves. We are still the only ape which shares food!

A successful group would have been the group which understood the basic rule of treating other people the way you would want them to treat you, and the most successful groups, by definition, would be the groups which left the most descendants. The most successful groups would have been the ones who understood that you succeed by doing least harm and that generosity pays off eventually. The most successful groups would have been those groups who cared for one another and gave a helping hand when one was needed. The most successful groups would have been the altruistic ones because altruism produces more survivors who carry the genes for, or the cultural idea of, altruism.

In short, the most successful groups would have been the groups with morals and ethics, and they would have taken the land formerly occupied by those without them. Groups with morals would have been the survivors. The survivors were the ones who cared for one another.

Part of our growing up as a species and learning to learn would have been to evolve children who were receptive and easy to teach; willing to accept what their parents told them. The downside to this is that it has produced children who are easy to mislead and willing to believe things because they have been told they are true by authority figures. They then carry this misinformation and frequently deliberate disinformation into adulthood and teach it to their children. The evolution of this childhood gullibility is easy to understand in the context of our evolution in East Africa.

As our brains developed so our children needed longer and longer to develop before being capable of independent existence, yet part of acquiring life skills in growing children involves curiosity. Human children have perhaps the longest childhood of any species. Curious children learn about their environments, where food can be found, which trees are easiest to climb, where shelter may be found, etc., etc., etc. Unfortunately, overly curious children are also at risk from predators.

Children who were told to avoid the water-hole where crocodiles live, or the places where pythons lurk and who believe what they were told, would have lived to pass their gullibility on to their children; those who wanted evidence and went to look for themselves would have been quickly removed from the gene pool.

So, we evolved childhood gullibility as a survival strategy. The downside to it is that it has made us susceptible to all manner of superstition, including religion. It has also made children vulnerable to depredation by paedophile priests who exploit this evolved gullibility and deference to authority figures for their own selfish gratification. It also explains why almost all religions are inherited from parents and are not based on a dispassionate assessment of observable evidence. The same applies to cultural norms and to some extent ethics of course.

Humans are by no means the only species to have evolved intelligence and cultures but more than any other species they have benefitted from another major evolutionary advantage cultures gave them – the possibility of gene–meme co–evolution.[1]

Back now then to the evolutionary story of the human family – the *Homo*.

We have reached the point where the first hominid left Africa and struck out into a new world. This was the descendant of *H. habilis*, *H. erectus*. It is possible that another descendant of an early hominid also left Africa around that time and became *H. floresiensis*, of which a little more later on. It is also possible that another species also migrated out of Africa into Europe, *H. heidelbergensis*, and there are some authorities who think *H. sapiens* evolved in Eurasia and subsequently migrated back into Africa. From our perspective, the significant migration was that of *H. erectus*. The picture might be complicated by the number of different regional varieties that *H. erectus* gave rise to and how they interbred with later arrivals.

On the Dawkins' analogy, we have just reached that piece of your right middle–fingernail you normally clip off. We have come a long, long way!

There are basically three different views of *H. erectus*:

1. That *H. erectus* is an Asian species, distinct from the African *H. ergaster*, that gave rise to 'Java Man', Peking Man', etc.

2. That *H. erectus* is the same species as *H. ergaster* and hence the ancestor of *H. neanderthalensis* and the Denisovans in Eurasia and *H. sapiens*, possible via *H. heidelbergensis* and/or *H. antecessor*.

3. That *H. erectus* had a wide range of morphology, as shown by the thirty skulls found in the Dmanisi cave in Georgia, south of the Caucasus Mountains in South West Asia, and that *H. ergaster*, *H. erectus*, *H. rudolfensis* and maybe *H. antecessor* are merely variants of this single species, maybe African, European and various Asian subspecies.

It seems fairly certain that *H. erectus* left Africa around about 1.9 million years ago, the first humans to populate Eurasia and radiating into regional variants.

Into the Trees and Beyond!

End notes:——————————————————

[1] Just like genes, memes are replicators that are imperfectly replicated and passed into the next generation. The term was coined by Richard Dawkins in *The Selfish Gene* as a contraction of 'memory gene'. Like all replicators evolving in an environment which contains other replicators, they will readily form alliances that are mutually beneficial, and which result in more copies of themselves being present in the gene (meme) pool. A great deal of human history can be explained in terms of gene–meme co–evolution rather than simple genetic evolution. One example of a memeplex (cluster of memes with a particular collective function) is religion. In many ways, religions behave like a memetic virus in the human meme–pool analogous to the way viruses behave like a genetic virus in the species gene–pool.

12. Walking Out of Africa

There are a couple of possible routes out of East Africa into Europe and Arabia and thence into Asia. One is via the Nile Valley to the Mediterranean coast, across what would otherwise be the Sahara Desert; the second is across the Horn of Africa into what it now Yemen then up the eastern Red Sea coast to the Levant and across the Fertile Crescent north of Arabia to the Tigris and Euphrates rivers in Mesopotamia, or up the Indian Ocean coast of Arabia to Oman and thence into the Arabian Gulf, across the Strait of Hormuz and into Iran and India. From there routes are open to South East Asia and the East Indies via coast routes or via the North Indian river valleys of the Indus and Ganges, or up through the Caucasus into Central and Northern Eurasia.

There is also a third possibility; the so–called Saharan Pump. The Saharan Pump was a slow cycle of prolonged period of very wet and very dry weather. During the wet phase, the Sahara and Arabian deserts became savannah with lakes and rivers contiguous with the East African savannah and inhabited by plants and animals migrating from Africa. During the dry periods, the desert returned and even the Nile became a dry wadi as the water evaporated more quickly than it flowed into it. This dry period would isolate two populations either side of the Sahara to evolve by allopatric speciation[1], the northern population evolving into a Mediterranean/European climate. The Sahara would then be repopulated from Africa again as the rain, rivers, lakes and grass came back.

However they did it, *H. erectus* spread out of Africa and across Eurasia, down into the East Indies and maybe as far as Papua New Guinea and Melanesia. They turn up in the fossil record as 'Java Man', found at Trinil, East Java, in 1891; as 'Peking Man' found at Zhoukoudian in 1921, and at Dmanisi in Georgia, discovered between 1991 and 2005. The Georgian find of thirty skulls was originally given full species status as *H. georgicus* but has now been reclassified as a subspecies of *H. erectus*, *H. erectus georgicus*. The diversity of this group has prompted a reassessment of the *Homo* family tree with several former species now regarded as possible variants of *H. erectus*. *H. erectus* has also been found in Pakistan, India, and maybe Europe at Atapuerca, Spain.

Soon after *H. erectus* appeared in Europe it was followed by *H. antecessor* and *H. heidelbergensis*, if indeed these were different species. By about 850,000 – 950,000 years ago something with human feet was walking on the beach at Happisburgh, Norfolk, England and leaving their footprints in the mud. These were revealed by a severe storm in 2014 which removed the protective cliff and the remains of sand dune that had covered then shortly after five individuals made them. Fortunately, they were documented, photographed and moulds taken before being washed away. These footprints were made before modern humans had evolved in Africa, let alone having lived in Norfolk. (Ashton, et al., 2014)

The current consensus is that *H. heidelbergensis* was the immediate ancestor of both *H. sapiens* and *H. neanderthalensis*, the former evolving in east Africa and the latter from the *H. heidelbergensis* that migrated up into Eurasia. Quite where *H. antecessor* fits in this scheme, the only examples of them being found in Atapuerca, Spain in the *Sima de los Huesos* (Pit of Bones), is uncertain. Some authorities regard *H. antecessor* as yet another subspecies of *H. erectus*, *H. erectus antecessor*.

This large collection of hominid remains is still being studied but what is clear is that they have characteristics of *H. erectus*, *H. heidelbergensis* and *H. neanderthalensis*. They are assumed to be intermediate between *H. heidelbergensis* (or a European variety – subspecies? – of *H. erectus*) and Neanderthals. The *Sima de los Huesos* specimens have been reliably dated to between 1.2 million and 600,000 years ago. They are the oldest fossil hominid remains so far found in Western Europe – which makes the Happisburgh footprints from 850,000 – 950,000 years ago all the more tantalising and intriguing.

Whether or not the *Sima de los Huesos* fossils were *H. erectus antecessor*, *H. antecessor* or *H. heidelbergensis*, there is little doubt that they were the immediate ancestors of *H. neanderthalensis* – the Neanderthals and probably of the Denisovans.

Another model has *H. heidelbergensis* being intermediate between the Eurasia *H. erectus* population and *H. rhodesiensis* being intermediate between the African *H. erectus* population and *H. sapiens*. This moves our last common ancestor back one to *H. erectus* and makes us a cousin species to Neanderthals rather than a sister species. But this debate is beginning to resemble a debate about how many angels can dance on the head of a pin. It is academically interesting but no more. Any confusion is more likely to be a confusion of classifications. There is no doubt that both modern humans and Neanderthals had their origins either in Africa or in species that originated in Africa. *H. sapiens* and *H. neanderthalensis* are very close, sharing 99.7% of their DNA, compared to both sharing 98.5% of their DNA with chimpanzees.

Probably the most compelling argument for an African origin of all modern humans is that there is far more genetic diversity in Africa than in the entire rest of the world. Genetic diversity is a measure of the age of a population. The lack of genetic diversity in extra–African modern humans is strongly indicative that the rest of the world was populated from a small number of migrants.

It is interesting too that in the 19th Century debate over the African or Asian origins of modern humans, with one side pointing to 'Java Man' as evidence of an Asian origin, Charles Darwin came out on the side of an African origin based on the fact that our closest relatives (based on anatomy) were the African apes. He argued that this suggested humans were another African ape. Le Gros Clarke later also came out strongly for an African origin – a view he felt was vindicated by the Leakey's discoveries in Olduvai Gorge with the fossil skull I narrowly missed seeing as a thirteen year–old.

There is little doubt now that the Neanderthals were widespread across Western Europe, the Middle East and Western Asia and that they were a distinct species to modern humans. They were almost certainly a Eurasian species that evolved *in situ* probably from the same ancestor as modern humans who evolved in situ in East Africa. The usual suspect for the last common ancestor is *H. heidelbergensis*, a descendant of *H. erectus* or maybe *H. ergaster* or *H. rudolfensis*. Some *H. heidelbergensis* had remained in Africa while some had migrated into Eurasia. Just possibly, modern humans evolved in Eurasia and migrated back into Africa, although the greater genetic variability of modern humans in Africa argues strongly against that idea.

Despite eventually becoming extinct, Neanderthals were a very successful species who lived in Eurasia for about 200,000 years – compared to modern human occupancy for only about 40,000 years so far. Their fossil remains become very much more common after about 130,000 years ago. They were adapted to life in a northern climate, having relatively short legs and a larger body than *H. sapiens*, giving a lower surface area to volume ratio, so reducing heat loss. Neanderthals had probably evolved adaptive skin colour and type and an immune system adapted to local pathogens. These were to be especially useful to modern humans when they came out of Africa and came into contact with Neanderthals, probably first in the Middle East and later across Eurasia as *H. sapiens* spread into Neanderthal territory. They interbred, not often, but often enough for modern humans to inherit some of the genes that Neanderthals had taken maybe 250,000 years to evolve. They gave the non–African population of *H. sapiens* a short cut to evolution into the much harsher conditions of the northern climates.

The evidence for interbreeding was first published in May 2010 (Fu, et al., 2014) and estimates put the amount of Neanderthal DNA in Europeans at 1–4% but this is unevenly distributed in the genome, being much higher in genes involved in skin proteins such as keratin and genes involved the immune system. This suggests that beneficial Neanderthal DNA was retained while deleterious DNA was quickly lost.

The negative side to this mostly beneficial transfer of Neanderthal DNA to *H. sapiens* is that it Neanderthal DNA has been implicated in autoimmune diseases such as arthritis, lupus, Crone's disease, Type 2 diabetes and even some mental illness. All of these are normally diseases of later life so there is low selection pressure to eliminate them.

It is especially noticeable that no Neanderthal Y chromosomes have been found in modern humans, as would be expected of people with a male Neanderthal ancestor. This was explained by the finding in 2016 that Neanderthal and *H. sapiens* Y chromosomes has diverged to the point where incompatibility cause *H. sapiens* women to abort a foetus with a Neanderthal Y chromosome (Mendez, Poznik, Castellano, & Bustamante, 2016). It is also noticeable that no Neanderthal mitochondrial DNA (mtDNA) has even been demonstrated in modern humans. mtDNA is inherited exclusively from the mother. The assumption is that fertile interbreeding would have been exclusively between Neanderthal males and *H. sapiens* females although mathematical models have shown that there would be a 93% probability of Neanderthal mtDNA going extinct by genetic drift alone and only a 7% probability of it surviving into today's human population.

If successful interbreeding between males and females of both species and if there was differential viability of any hybrid combination, then this argues strongly against Neanderthals and *H. sapiens* being simple variants or even subspecies of the same species, and in favour of them being different species which could still interbreed to a limited extent.

Anthropologists believe that Neanderthals had a low population density and lived in small, isolated groups. They were apex predators living on large game that they hunted, supplemented with wild plants. There is no evidence of cultivation or domestication of animals, probably not even dogs. In fact, the domestication of dogs has been proposed as one of the reasons modern humans were more successful ultimately and replaced Neanderthals over their entire range in a period of only about 3–5,000 years gradually being reduced to their last strongholds in Southern Britain, Brittany in France and the Iberian Peninsula. The last remaining Neanderthals are believed to have lived in Gibraltar where they may have survived until about 23,000 years ago. (Finlayson, et al., 2006)

So, what did the Neanderthals do for us?

Very little for the African populations, for whom there is little evidence of Neanderthal genes, but for the northern groups, one of the things they could have given us was pale skin. The reason this was beneficial is because of the relationship between skin cancers caused by excessive ultraviolet (UV) radiation in sunlight and the need to manufacture Vitamin D in our skins by the action of sunlight to supplement a deficiency of it in our diet.

UV radiation in sunlight can cause a mutation in the skin cells, causing them to become the rapidly–disseminating skin cancer, melanoma, but this is normally filtered out by the skin pigment, melanin. The reason most populations, other than recent migrants, in tropical zones are dark skinned is because this protective pigment, almost certainly evolved as we lost body hair was inherited from ancestral species. It is thought that the first modern humans that left Africa were almost certainly dark–skinned.

But what was a protective pigment in tropical climates proved to be a problem in norther, cloudy climates where sunlight falls more obliquely on the Earth. Dark–skinned people in cloudy northern climates do not manufacture enough Vitamin D in their skin. Vitamin D synthesis is triggered by UV radiation and is essential for normal long–bone development in children. Deficiency in Vitamin D caused the deformity once common in Europe, rickets.

So, the answer was to lose most of the melanin pigment. The trade–off was that it made the Eurasians more susceptible to melanoma (which is rare in Africans except occasionally on the soles of feet and under the fingernails, where there is little or no melanin. It is believed Europeans inherited this earlier adaptation to northern climates from Neanderthals, not directly but from a Central Asian people who migrated into Europe, bringing pale skin with them. The genes for pale skin then spread rapidly through the population, showing that Vitamin D deficiency might have been a significant problem.

Needing to balance the risk of melanoma against the risk of Vitamin D deficiency is surely another example of the mindless, unplanned and unintelligent nature of evolution. Even a half–competent intelligent designer should have been able to produce something better than that.

This was almost certainly not the first time that *H. sapiens* or their ancestors had interbred with other close relatives. In 2010 archaeologists discovered a 40,000 year–old finger bone of a juvenile female in the Denisova cave in the Altai Mountains in Siberian Central Asia. Analysis of the mtDNA from this bone showed that it was from a species distinct from both Neanderthals and *H. sapiens*. It was from an entirely new archaic human. Pending a definitive status it has been designated *Homo sp. Altai* or *Homo sapiens ssp. Denisova*. They are colloquially known as the Denisovans.

Subsequent nuclear DNA analysis has shown that they were close to but distinct from the Neanderthals and probably split from them about 640,000 years ago and that they both shared a last common ancestor with modern humans 804,000 years ago.

The Denisova cave was also occupied by Neanderthals, the remains of which have also yielded DNA showing that the two populations interbreed with local Neanderthals having about 17% Denisovan DNA. Denisovan DNA is also found in modern humans and especially in native Australian and Melanesian people, suggesting that Denisovans were probably distributed across East and South East Asia and interbred with *H. sapiens* as they migrated into the area.

There is also evidence that Denisovans interbred with another, so far unidentified, hominid. Somewhere in Eurasia there was at least one more species or subspecies of humans whose remains have yet to be discovered but who left some of their DNA in the Denisovans. Was this *H. erectus* or an offshoot, or an ancestor species of *H. floresiensis*, or some other species? Maybe we'll never know. Certainly, a lot more Denisovan DNA would be useful and would help give this ancestral species a proper name and a definite place in the family tree.

One of the fossil remains of a female found in *Sima de los Huesos* in Spain has a mixture of Neanderthal and Denisovan DNA but more Denisovan than Neanderthal, suggesting again an interbreeding population (Callaway, 2013). One curiosity is that of the two teeth from different individuals found in the Denisova cave, the mtDNA recovered shows the Denisovans had a wide diversity of mtDNA, more than the entire Eurasian population of Neanderthals so far sequenced and wider than modern humans from different continents.

In 2014 a team of researchers found that the adaptation to high–altitude living of Tibetans was not entirely due to mutant genes they had inherited from their recent Han Chinese ancestors but also due to a variant gene known as *EPAS1* which probably spread very rapidly through the population when they acquired it from their Nepalese Sherpa neighbours. This variant is unknown in any other modern human population but is identical to one found in Denisovans! (Huerta-Sanchez, et al., 2014)

It is clear that for much of our history, humans existed as a number of different, contemporaneous species, more or less geographically isolates but occasionally interbreeding with greater of lesser success. This is exactly what you would expect of a wide–ranging population existing mostly as isolated groups; a genus in the process of diversifying with occasional gene–flow between populations. It is perhaps more of a mystery where the other species went than where they came from. It looks suspiciously as though modern humans, *H. sapiens*, exterminated the others, the last ones maybe as recently as 23,000 years ago in southern Spain.

The result was our species, the sole representative of a genus that evolved in East or South Africa and now present in every continent except Antarctica and regarding itself as ruler of Earth and owner of everything on it. *Homo sapiens,* the thinking Man, had won the evolutionary struggle and was now guardian of your genes.

End notes:——————————————————

[1] Allopatric speciation is caused when two populations become separated by a barrier to interbreeding. Each population is then free to evolve according to local conditions. If isolated for long enough they may progress to different species, unable to interbreed if they come back into contact.

13. The Final Steps

So, we enter the closing chapter of your long journey through space and time. A history that has seen you go from as near to nothing as it can possibly be to energy in the form of gravity, electromagnetic radiation and weak and strong nuclear forces, to atoms of light and heavier elements. We have traced your history in suns and supernovae and accretion discs and planets and we have seen how natural forces conspired in the right conditions to create self–replicating molecules and eventually, given enough time, to free–living, self–replicating entropy management machines we call life with the first genes that are still found in your cells.

We have seen how your ancestors learned to make sugar from sunlight and to split water to create Oxygen, not because it was planned but because it worked and because the billion to one chances come up regularly given a large population and lots of time. Then we saw how your ancestors evolved under nothing more than the selection pressure of their environments to make multicellular organisms, animals with backbones, and fish with lobed fins that could use them for walking. We have seen how your ancestors learned to gulp air to get the oxygen out of it and how this made them able to leave the water to catch the first land animals.

We have followed your ancestors through the steaming Carboniferous and saw how they survived the Carboniferous Forest Collapse to inherit the land as reptiles and the beginnings of mammals. We have seen how the dinosaurs made the night time your ancestors' preferred time and how this helped them survive the K–T extinction and years of permanent winter. Then we saw how your ancestors learned to retain their eggs and maybe with the help of an ancient endogenous retrovirus, to become placental mammals, nourishing their young inside their mothers' bodies and feeding them on breast milk.

We have followed your ancestors into the trees and back down again into the savannahs of East and South Africa. We have seen them learn to walk upright, grow a big brain and learn to learn and develop cultures and ethics and cooperative social groups. We have seen them diverge and spread and interbreed and populate the world.

And we have seen your ancestors win the evolutionary struggle with other humans and inherit the earth.

Now we can finally tell the story of how your ancestors spread ideas and technologies and went, in many parts of the world from bands of hunter–gatherers living a feral existence to settled agriculturalists, nomadic pastoralists, urban citizens and consumers of industrial production able communicate across the world in the blink of an eye, and to write, print, buy and read books like this explaining how it all came about.

Almost all of what follows would be wiped away with a single stroke of a nail–file on the middle finger of our outstretched right arm.

The course of human history did not follow the same path everywhere. The course that your personal history took will depend on where your parents originated. It will almost certainly be a mixture of one or more of these. For example, I was born in England to English parents whose ancestors as far back as I have traced them, lived almost wholly in Oxfordshire, especially the Wychwood Forest area, Northamptonshire and Buckinghamshire. The earliest ancestor I have been able to trace was a man who lived in 1550 in a village 2 miles from where I was born, yet my DNA analysis shows that I am 28% Scandinavian, 21% British, 18% Irish, 15% Western European and 11% Iberian. The other 7% is from Italy/Greece, Finland & Northwest Russia and Eastern Europe. One of my father's cousins is a sickle–cell gene carrier and both my granddaughters have the 'D' version of haemoglobin.

My partner, whose maternal grandparents were from County Mayo, Ireland and whose father's parents and grandparents were from Norfolk, Northamptonshire, Somerset and Sussex, is 61% Irish, 16% Scandinavian, 9% Western European, 6% Iberian and only 4% British. The rest is 2% Italian/Greek and the other 2% a mixture of Finish/Northwest Russian, and Middle Eastern. Although British to the core, neither of us is predominantly British, according to our DNA.

Try this as an exercise: write down your country of birth, the country of birth of each of your parents, each of your grandparents and each of your great grandparents. Most people, especially from the Americas, Australia, New Zealand and Canada and much of Europe will have at least two or more countries in that list and often two or more continents. The fact is that we are not, and have not been for a very long time, members of more or less isolated genetic groups. We are all of us a mixture of human groups, even if we think we are not. Use your list to trace your probable cultural and genetic ancestry from the following brief histories.

Africa.

The birthplace of mankind and the crucible of the first technology – the shaping of wood, bone and stone into tools and the discovery of fire, Africa lacks two important ingredients: large animals suitable for domestication as beasts of burden or food and suitable plants for cultivation as a staple diet.

Africa also suffers from two major problems:

1. The tsetse fly which debilitates imported domesticated animals and makes it impossible to use anything more than manpower for tilling the land and transporting goods. Using manpower alone it is just about possible, in a suitable climate for a man to work enough soil to feed himself and his family. Producing a surplus to sell for goods and services is difficult or impossible and without a surplus to feed the producers of goods and services in towns, urban development and specialisation of labour into crafts will not happen. The only part of Africa south of the Sahara to have achieved substantial agriculture and an urban way of life was Ethiopia in the highlands where the tsetse fly is absent.

2. Unreliable rainfall. Prolonged periods of drought and the absence of major rivers and lakes to function as reservoirs for irrigation has made agriculture impossible even when suitable sources of energy other than manpower have been available. Much of the Southwest and North are arid deserts and parts of the East are subject to prolonged periods of drought.

Without the surplus of food production, the only alternatives are subsistence farming and hunter–gathering. It is not possible to develop urbanisation, political organisation, taxation to finance government nor to adopt and use innovative technologies like bronze and iron smelting and the manufacture of fabricated goods. For this reason, Africans, although the most genetically diverse, not having gone through the genetic bottlenecks of occasional migration of small bands, remained in sub–Saharan Africa, with some notable exceptions, as either Iron Age subsistence farmers, often using imported iron, or Stone Age hunter gatherers.

Gold, mined in Ghana and Mali was the stimulus for trans–Saharan trade with North Africa, made possible by the introduction of camels from the Middle East in the 5th Century CE. Trade in the other direction consisted mostly of salt. Control of the Bambuk gold fields gave power to the Soninke who established the ancient kingdom of Ghana and exerted control over the Malinke. The kingdom of Ghana eventually succumbed to attacks from the Mossi, a people from Southern Ghana and the Volta River valley. This allowed the rising Songhai Empire to gain control of the gold fields.

West Africa appears to have skipped the Copper and Bronze Ages and gone straight into the Iron Age, as long ago as 2000 BCE. A 2009 article by Heather Pringle in *Science* (Pringle, 2009) summaries the competing arguments about when and how Africa acquired iron smelting technology. Some claim evidence that it was discovered independently while others claim it was imported across the Sahara by North African traders.

There are many African myths and legends about ancient people with magic skills able to turn rocks into metal, one such being the legend that Rwanda's ancient line of kings was descended from one such man with magical knowledge and powers. One explanation for the explosive radiation of Bantu people from West Africa into East and South Africa, pushing the San people into Botswana and the Kalahari, is that Iron Age technology cause a rapid increase in populations.

The Iron Age in Southern Africa briefly gave rise to the substantial kingdom of Zimbabwe, probably built by the Gokomere people, the reputed ancestors of the Shona. At its height, the capital city of Great Zimbabwe, the only substantial stone buildings in Africa outside Egypt, held as many as 18,000 people. It was built between the 11th and 15th Centuries (Beach, 1998). Zimbabwe is the Shona name for the ruins.

From artefacts excavated at Great Zimbabwe, including shards of Chinese pottery, it is clear that there was an extensive trading network, possibly with Arab traders and others, either overland or via the coast in the vicinity of Beira, probably based on gold from gold mines between the Zambesi and Limpopo Rivers. When these mines became exhausted, the Kingdom of Zimbabwe went into decline.

The San people are anthropologically interesting in that an analysis of both their Y chromosomes and mtDNA shows them to be the most diverse of any of the 121 different African groups, suggesting they have been isolated for a considerable period of their history. The evidence of fully sequenced genomes shows that the San began to diversity from other humans about 200,000 years ago and were fully isolated by 100,000 years ago, hence they are probably closest to the ancestral *H. sapiens* of all human groups. 44,000 year–old tools identical to those used by modern San hunters have been recovered from a cave in KwaZulu–Natal.

The lack of political organisation in Sub–Saharan Africa made the continent easy pickings for European colonial powers looking for slaves and minerals.

Africa north of the Sahara was a different matter entirely. The Nile, and its delta, like other major river valleys in other parts of the world, was to provide the basis of settled productive agriculture based on the cultivation of barley and emmer (an early form of wheat), cities and a thriving culture, possibly the first, in Egypt. Originally two separate states in Upper and Lower Egypt respectively, Egypt became unified as a single political entity traditionally in about 5350 BP. Egyptian culture spread into the Mediterranean and into the Levant, down into Northern Sudan and across the Red Sea into Western Arabia and probably modern Yemen.

It is by no means certain who the ancient Egyptians were – indigenous Africans or migrants returning to Africa from the Middle East. Some scholars think it that the Southern Kingdom based on Thebes was African while that in the Delta was either from the North African Maghreb or the Middle East. They may have been derived from a Saharan people driven into the Nile valley as the Sahara became a desiccated desert.

Even today, a traveller down the Nile from Luxor (ancient Thebes) to Aswan will notice the local Egyptians becoming more African–looking the further south they go.

Ancient Egypt fell under first Greek and then Roman political and cultural influences and, as these powers waned, the influence of expanding Arabic political, religious and cultural power from Arabia, Syria and Iraq in the Early Middle Ages.

One part of geographical Africa that is noteworthy by its difference is the island of Madagascar. Not only is much of its flora and fauna different to mainland Africa but the people of Madagascar are not even African. Africans do not appear to have developed much in the way of sea–faring capability and certainly ever reached Madagascar.

Having split off from the African mainland along with India when the ancient landmass, Gondwana, split up around 135 million years ago, with India later splitting off and going its own way, Madagascar has been biologically isolated ever since. 90% of Madagascar's flora and fauna is found nowhere else on earth.

As linguistic and genetic evidence shows, Madagascar was peopled by Polynesians using their wind–powered outrigger canoes with which they populated the Pacific. Riding trade winds and ocean currents across the Indian Ocean from Southern Borneo, this must rate as one of the most amazing migrations of any people in human history. They appear to have arrived in successive waves between 350 BCE and 500 CE, although some claims have been made for visits as long ago as 2000 BCE. Madagascar is thus one of the last major landmasses to be colonised by humans.

Southeast Asia.

As with Egypt, settled agriculture, this time based on the cultivation of rice as the staple diet, developed in major river valleys in China, the main ones were the Yellow River (Hwang Ho) and the Yangtze. Agriculture led to towns and the production of goods for sale in exchange for crops from the surrounding countryside.

The co–evolution of rice and the humans who cultivated it is an interesting example of how the evolution of two disparate species can become inextricably linked. It had been generally assumed that rice cultivation began in China along the Yangtze River Valley some 9,000-10,000 years ago when humans first domesticated wild rice *Oryza sativa* of the *japonica* variety. This cultivar then spread throughout Asia and into the Indian subcontinent where it hybridized with other wild varieties to give, for example, Basmati aromatic rice.

This view was questioned by a team of researchers led by Peter Cìván of the University of Manchester, UK, (Cìván, Craig, Cox, & Brown, 2015) who believe they have shown that, although *japonica* rice was cultivated in China, so were two other varieties independently in the Indus Valley in modern Pakistan where the *indica* variety was cultivated, and in what is now India and Bangladesh along the Brahmaputra River system where the *aus* variety was cultivated. Both of these river valleys gave rise to flourishing early civilisations as humans learned agriculture and changed from hunter-gatherers to settle agriculturalists.

The researchers arrived at this conclusion having examined the genomes of 1083 varieties of modern rice and 446 samples of wild rice from all over southern Asia. It contrasts with a 2011 conclusion by a New York University team who concluded that rice had been domesticated just once, in China. Typical of scientific debate however, both 'sides' acknowledge that these different conclusions could be due to the regions of DNA chosen for the study and that neither answer is definitive.

Biologically, humans created the conditions in which the genes of the different varieties of rice could form an alliance with those of humans and so both flourished in this alliance. Both humans and rice are now vastly more numerous and biologically successful than they would have been otherwise.

Exactly the same point can be made for other domesticated plants and animals, of course. It is recognition of the power of 'selfish' genes to act in mutually self-interested alliances that gives the lie to claims that 'selfish gene' evolution can never give rise to operation and altruism. In fact, nothing could be further from the truth.

By being spread geographically by humans, these different varieties were brought into close proximity, having diverged into geographical varieties earlier in their evolutionary history. When this gave rise to cultivars which humans preferred, such as the wonderfully savoury aromatic Basmati rice - a hybrid between *indica* and *aus* - or the conveniently 'sticky' *japonica* rice where chopsticks are used, or the round, absorbent varieties used for puddings and paellas, these were selected for and so the rice plant evolved into today's multiple varieties.

If people would cease the religiously inspired, nonsensically arrogant assumption that humans are somehow apart from nature, selective breeding like this would be seen as merely a form of natural selection and just as much an evolutionary process as any other adaptive process. From the rice plant's 'point of view', this evolutionary change can be seen as rice genes co-opting humans to aid in their propagation and persistence over time.

An example of how environments drive evolution by providing the natural selection of those best able to reproduce from among the different varieties can be seen in the distribution of an allele which gained traction when humans began to cultivate rice in China.

It is quite common amongst Asian peoples that shortly after drinking even a small quantity of alcohol, they display an 'alcohol flush', rather like a facial blush. People who do this appear to be able to tolerate alcohol better than others. Researchers at the Chinese Academy of Sciences (Peng, et al., 2010) have discovered that this is caused by a variant allele called *ADH1B*47His*. Molecular dating techniques have shown that this arose in China around 7-10,000 years ago, exactly when Chinese culture underwent a major (possible the most significant) change when rice was domesticated. This crop quickly became the staple source of carbohydrate and a good deal of the protein in the Chinese diet and enabled settled agriculture and the growth of cities in and around the Hwang Ho valley.

Rice can also be fermented to give alcohol which has several uses: It can be used recreationally but ultimately destructively when used in excess; it can also be used to preserve food and enhance its nutrient qualities, as a disinfectant and medicinally as an analgesic. It is believed that drunkenness may have quickly become a major problem giving both increased food but also increased problems with drunkenness and alcohol-related illness. So, people carrying *ADH1B*47His* would have been able to tolerate alcohol and would have suffered less damage from recreational alcohol, whilst benefiting from the positive benefits, giving them a very real advantage.

A distribution map of the occurrence of this allele corresponds closely with the distribution of rice growing, and the incidence is correlated with the length of time rice has been in cultivation in the area. So here again we see an illustration of the way the environment drives evolution by translating the information in the genome and giving it meaning. Before there was rice being cultivated and the alcohol that was able to be produced by it and with it, the *ADH1B*47His* allele had no meaning whatsoever. In the presence of rice and alcohol it meant survive and prosper when others are suffering and failing. And so *ADH1B*47His* increased in the local human gene pool. The presence of rice changed the environment of the human population of China and evolution ensured the human population promptly adjusted to fit into this new environment.

With urbanization in China came political organisation and nationhood. The surrounding peoples in Korea, Japan, Tibet, Vietnam and north into Manchuria became Sinicised acknowledging the Emperor to a greater or lesser extent. China has been remarkably culturally if not politically stable for much of its history. Even 'barbarian' invaders from the north (Mongols and Manchurians) quickly became Sinicised. When independently–minded Japan declined to acknowledge the Mongol Emperor of China, Kublai Khan as emperor, this was considered sufficient grounds to launch a massive invasion. The spectacular failure of this invasion served to confirm the cultural and political independence of Japan.

One of the reasons for China's political stability, despite the many mutually unintelligible dialects spoken in China is because of Chinese writing. Chinese characters express ideas rather than sounds and so can be read with equal ease in Mandarin, Cantonese, Wu, Min, Xiang, etc. But even typing a letter in Chinese is fraught with difficulties.

China made some significant technological advances but never invented printing (and so mass dissemination of information), probably because of the complexities of the huge Chinese pictogram alphabet with thousands of characters. One of the reasons China did not progress technologically as quickly as Europe was that they lacked the technology to make clear glass, hence they could not make microscopes, telescoped, spectacle to extend the working lives of scholars and suitable chemically inert containers for experiments in chemistry. Of such things are differential cultural and technological progress made.

Central and Northern Asia.

The vast steppes of Central Asia are generally arid and, in the winter, bitterly cold and unsuitable for agriculture, so these areas remained mostly nomadic hunter–gatherers or migratory pastoralists. Their distance from the main population centres in the East and West and from the coast, which is subject to freezing anyway, made trade difficult. The predominant trade which did develop was the silk trade from China to Europe and the Middle East and the movement of goods in the other direction to exchange for silk. This trade depended on a degree of political control and stability, such as that provided by the Mongol Empire under Genghis Khan, to control bandit attacks and make the long caravan journeys safe.

Society was highly tribal and dominated by horse–mounted expert bowmen. Changing tribal loyalties would sometimes produce a powerful single military force able to dominate and coerce others into joining in devastating conquests into the civilised states in the east and west. Such tribes as the Huns, the Turks and the Mongols all had their impact on Eurasia, the latter establishing a dynasty that stretched from Vienna to Beijing and was able to mount an invasion of Japan and threaten Jerusalem – a threat that was only lifted when Genghis Khan died and the military commanders returned to the Mongolian homeland to decide on the succession.

Genghis Khan is interesting in what we can learn about human evolution from the way his genes spread throughout Eurasia by a process of gene–meme co–evolution.

In humans, genes can increase in frequency in a population not necessarily because they represent an adaptive advantage in their own right but because they are bound to other replicators, such as cultural memes, which give them an advantage. This linkage may be entirely due to chance. For example, a culture which is hierarchical and expansionist, and especially where sons inherit the power, authority and privilege of their fathers, may facilitate the spread of genes carried by powerful men, especially where power gives access to females and comes with a higher standard of living so children are more likely to survive. The actual genes benefiting from this may be completely unremarkable.

A paper published in the European Journal of Human Genetics in 2015 (Balaresque, et al., 2015) showed that it was possible to identify a number of clusters of a particular haplotype of the Y chromosome (only carried by males and so indicating the male inheritance line). By counting the number of mutations in a given region of the Y chromosome and assuming a regular mutation rate, it was possible to estimate the time when this variant arose in human evolutionary history. By assuming that the geographical location where most diversity was found was close to the place where the variant originated it was possible to estimate a likely place where this variant arose.

Two previous clusters have been identified and associated with Genghis Khan, the Mongolian leader who founded the vast central Asian Mongolian Empire which stretched from Austria to China and who died in 1227, and a Chinese Qinq Dynasty Emperor named Giocangga who died in 1582. Another cluster has been found associated with a legendary Irish tribal leader, Niall of the Nine Hostages, putative founder of the Uí Néill, probably the major power in pre-Christian Ireland and for a considerable time after.

The secret to getting your Y chromosome into future generations is, of course, to have lots of sons, and for those sons to have lots of sons, preferably for three or more generations, and what better way to do that than to be a powerful tribal leader? Genghis Khan reputedly had an organised band whose job it was to keep him supplied with several young women a night and he reputedly fathered hundreds of children. His sons by his official wife, Jochi, Chagatai, Ögedei and Tolui all became powerful rulers of khanates after the death of Genghis.

No doubt his many 'unofficial' sons would have been high status too. His four sons by his Empress Börte all had several sons many of whom also became powerful rulers. Of Tolui's many sons, Möngke became Great Khan of the Mongol Empire, Kublai became the first Yuan Dynasty Emperor of China as well as Great Khan of the Mongols, and Hulagu became the first Ilkhan of Persia.

Not surprisingly then, 8% of men in 16 populations in Asia and 0.5% of men worldwide have Genghis's Y chromosome. It is difficult to imagine any other Asian ruler of that time able to command the social and economic power to produce as many surviving and powerful male descendants as Genghis Khan, so there can be few serious contenders for the progenitor of this particular Y chromosome cluster. It is always possible that Genghis inherited his unique Y chromosome from his father or grandfather and that the key mutation didn't arise in the sperm that contained his Y-chromosome but it is surely to Genghis, his sons and grandsons that credit must go for its disproportionate success rate.

In the case of Giocangga, it is estimated that he had about 1.5 million descendants in 2005, mostly in and around northern China and Mongolia where this Y chromosome cluster is found, whereas it is almost completely absent in Han Chinese. The Qinq dynasty was founded by upstart Manchurian tribal warlords from the north, ruled for many generations and the emperors normally had many concubines. Again, we see several generations of powerful men with many sons, many of whom were also powerful men in their own right, and a political process which continued this for several generations.

In the case of Niall of the Nine Hostages (*Niall Noígíallach*) little is known of the historical person but the *Uí Néill* super-dynasty he founded were powerful 'High Kings' of Tara, nominal overlords of dozens of petty kings or tribal chiefs in Ireland, from the 6th to the 10th Century CE. The *Uí Néill* family of smaller dynasties or *Cenél* all trace their origins back to the eight sons of *Niall Noígíallach*. One of the northern branches went on to found the kingdom of *Dál Riata* on the north-east coast of Ulster which extended across the Irish Sea to western Scotland, taking Gaelic and the Irish 'Scoti' (a sept of the *Uí Néill*) into Scotland.

Of his great-great-grandsons one was *Colm Cille*, or St Columba who took Christianity to Scotland, converting the Picts and facilitating a union of the Gaelic and Pictish tribes to found modern Scotland. Another great-great-grandson was Saint *Máel Ruba*. One of the last Gaels to play a key role in Irish history was Hugh O'Neil, Earl of Tyrone, on whose confiscated lands following the 'Flight of the Earls' from Ireland, the Ulster Protestant Plantations were founded. The rebellion Hugh O'Neil led was the last attempt by the native Irish Gaels to defeat the English and regain Irish independence. Hugh O'Neill was a direct descendant of *Niall Noigiallach*.

Irish geneticists have estimated that 21% of men in north-western Ireland, 8% from the rest of Ireland and 2% of New York men carry Niall's Y chromosome. Chances are that if you are male and an O'Neil or one of its variants, you do to.

What we have in all these situations is an ability to produce a lot of male offspring who also produce a lot of male offspring - the classical recipe for evolutionary success, in this case of the 'male' chromosome. But in these cases, of course, the thing that allowed these males to produce lots of males and so to give their Y chromosome an advantage over other Y chromosomes, had little or nothing to do with genes. It was almost all to do with the other major replicators which play a large part in human evolution - cultural memes. It is probably no coincidence that a lot of the estimated times these clusters originated and the places they originated in, are in areas where political and economic development would have produced a hierarchical society.

So, we see here an example of how memetic culture can form an incidental alliance with genes and cause these genes to become much more numerous in the gene pool for reasons having little or nothing to do with the genes themselves. Human cultural hierarchies, inherited power, and the access to females that that often brings about, are of course the biological equivalent of the alpha or dominant male with a harem such as we see in many herding and social species. It may be that some genes are involved in this ability to dominate but often, especially with humans, the main factor might be who your father was rather than what genes you got from him or your mother, and especially what your culture is if that favours inherited power and deference to authority. There will be many other genes riding piggyback on any genes involved just as there will be many riding piggyback on the cultural memes.

The nomadic lifestyle which had dominated the steppes of Central Asia was brought under control in the 16th Century when the firearm gave the civilised world the ability to defeat the nomads and bring the area under control.

In prehistory, Central Asia, especially the Lake Baikal area, was probably the homeland of the original inhabitants of the USA (Rasmussen, et al., 2014)and there are close affinities between some coastal Siberian people and the Inuit and the Aleut people of Alaska and the Aleutian Islands. There is evidence too that the people who migrated to the Americas from the Lake Baikal area, almost certainly via the now lost land of Beringia, between Alaska and Siberia but maybe over sea ice, also migrated westward and contributed to the modern Europeans. (Lazaridis, et al., 2014)

South Asia.

There is evidence of *H. erectus* living in South Asia as long ago as 1.9 million years ago at Riwat in Punja (Chauhan, 2003) and of *H. sapiens* activity as long as 75,000 years ago. There are rock paintings at the Bhimbetka rock shelters dating from about 30,000 years ago. The first signs of agriculture began to appear from 9000 BCE, replacing stone–age hunter–gathers. Opinions differ on whether agriculture developed independently or was imported from the Middle East via Persia – or a mixture of both. Cultivation of rice may have spread from China and was based on the *indica* variety and Basmati – a hybrid of *indica* and the Chinese *japonica* variety.

As with Egypt and China, settled agriculture and urban living developed in river valleys. From about 2300 BCE the Indus Valley was home to the Hrappan civilisation named after the first city excavated in the 1920s. A second city was found at Mohenjo–Daro soon afterwards. At its height, the Indus Valley civilisation contained an estimate seven million people. When the Indus Valley civilisation collapsed, possibly as a result of desiccation caused by climate change, in about 1750 BCE the area was colonised by Indo–Aryan pastoralist migrants from Eurasia. Linguistic analysis shows a common ancestral language with other Indo–European languages. The Indo–Aryan Vedic religion was the progenitor of modern Hinduism, Jainism and Buddhism.

Australia, Melanesia and the Pacific Islands.

It is generally recognised that Australia was reached fairly early on by the first modern humans to leave Africa and the assumption is that this first wave of migration was by coastal spread. The earliest *H. sapiens* remains to be found in Australia, near Lake Mungo in New South Wales are considered to be at least 40,000 years old. Genomic evidence suggests a date of around 60,000 years ago (Oppenheimer, 2003). Australian Aboriginals are known to have some

Denisovan DNA which they probably picked up during the initial migration through South East Asia.

There is conflicting evidence of whether or not there was an influx of migrants into Australia about 4000 years ago. On the one hand, is a recent analysis of thirteen Y chromosome of Australian Aboriginals, which show no signs of contact with other peoples since they diverged from other *H. sapiens* about 50,000 years ago. On the other hand, is the evidence that the dingo arrived in Australia about 4,000 years ago and almost certainly had its origins in the wild dogs of Thailand, suggesting at least cultural and trading contact with people from South Asia. The ancestors of dingoes are believed to be semi–domesticated versions of the Asian wolf and all have either the same A29 mitochondrial DNA, or one just a single mutation different. This suggests the entire Australian dingo population is descended from a single female (Oskarsson, et al., 2012) and came to Australia via the Malayan peninsula.

There is also conflicting evidence of agriculture and settlements including the suggestion that 'slash and burn' farming might have been practiced; however, the Australian Aboriginals were almost exclusively very low–tech Stone Age hunter–gatherers when they came into contact with Europeans. They did not have the bow and arrow but relied on a spear thrown with the aid of a woomera[1] to give it range and power, and on the boomerang.

The people of Melanesia, like the Australian Aboriginals, are believed to be descended from the first *H. sapiens* who migrated out of Africa and round the South Asian coast. Like them too they have some Denisovan DNA (3–4%) and Neanderthal DNA (2%), presumably picked up by occasional interbreeding as they migrated into South East Asia.

The Polynesian Islanders, including the Maori of New Zealand, have a different and more recent ancestry, however. Genetic studies have shown they have few connections with Melanesia Islanders and probably originated in Southeast Asia. The idea that they came from island of Taiwan prior to its colonisation by mainland Chinese has now been discounted following a more detailed analysis of mitochondrial DNA. This analysis suggests they originated from the Asian mainland but via a population that had already colonised islands close to Papua New Guinea by about 6–8,000 years ago (Soares, et al., 2011).

See also the migration of Polynesian people to Madagascar in the section above dealing with Africa.

Until they came into contact with the developed world, the peoples of Melanesia and Polynesia lived by cultivation of local crops, and by hunting, fishing and foraging on the sea shore.

The Americas and the Caribbean.

Before the arrival of Europeans, the people of the Caribbean were migrants from North, and possible South America. These people were themselves descendants of earlier migrations from Asia, almost certainly via a temporary land bridge between Siberia and Alaska, known by anthropologists and geologists as Beringia.

The Caribbean Islands were the first point of contact between Europeans and Native Americans in 1492, when, contrary to the myths, Columbus landed in the Caribbean and never set foot on the North American mainland. Nor, incidentally, did he set out to prove Earth was spherical. He knew perfectly well that it was, as did his backers in Portugal. The debate was over how big it was, and which was the shortest route to China. He assumed he had missed China by going too far south and had landed in India, hence the colloquial name of the Native Americans, 'Red Indians'.

The first people into North America were probably from the Lake Baikal area of Central Asia. One of the strongest pieces of evidence for this comes from the 13,000 year–old skeleton of a girl found in a flooded cave in Mexico. Although her skull suggested a relationship to people of South East Asia, Africa and even Australia, her mitochondrial DNA show she was descended from people from Siberia.

The skeleton of a three year–old boy, reliably dated to 12,600, buried with some considerable ceremony in the Rocky Mountains, had DNA that was shown to be equally close to that of Central and South American peoples and to have shared a common ancestor with Native Canadians. It was also close to that of a 24,000 boy from Mal'ta in the Lake Baikal area of Siberia. The conclusion is that his people were the ancestors of South and Central American people and that his people and Native Canadians shared a recent common ancestor from Siberia.

Linguistic analysis suggests three discrete waves of migration from East Asia but there is evidence that people spread very quickly down into Central and South America. A 2013 paper presented archaeological evidence that people were living in a rock shelter, the Toca da Tira Peia, Piauí, Brazil as long as 20,000 years ago (Lahaye, et al., 2013).

Very quickly humans spread throughout the Americas as hunter–gatherers, eventually forming settled communities based on agriculture and even building substantial settlements in the Peruvian highlands at Cuzco, the centre of the Inca Empire, despite the fact that they lacked animals for riding and draught, having only domesticated Llamas as pack animals. Inca society was probably highly feudal, with a slave economy and the granting of land for subsistence farming in return for feudal duties and 'tax' in the form of a share of the crops. There is no evidence of currency, so any trade would have been by barter.

Other groups diversified in Amazonia to form more or less isolated forest hunter–gatherer tribes, some of which may yet remain undiscovered.

In North America, the culture remained essentially nomadic hunter–gatherer with some agriculture in the south based on cultivation of *Zea mays* (Maize or Corn) as the staple diet. Maize is believed to have been first domesticated in Mexico about 10,000 years ago. It is an interesting example of co–evolution because cultivated maize cannot disperse its seed without human intervention because they do not fall from the enclosed ear. Exactly how and why it was first domesticated is still in dispute. The indications are that it was domesticated only once and may have been the chance result of a hybrid between two related species.

Cultivation of (primarily) maize supported a succession of civilisations capable of building a major stone–built settlement at what is now Mexico City, complete with the *Templo Mayo* (Spanish, Main Temple). The first of these cultures was the Olmec (1500–400 BCE) followed by the Toltec (c.900 – 1168 CE) and finally the Aztecs (1428–1521). In 1521, Spanish *Conquistadores* captured the town of Tenochtitlan, having massacred 600–1000 people, including the king, Moctezuma II, attending a religious festival in *Templo Mayo* the year before, effectively ending the Native American advanced civilisation in North America.

North America remained in the Stone Age until the conquest by Europeans from 1492 onwards which resulted in the extermination, often by deliberate genocide, of many native groups. Earlier contact with Vikings of Scandinavian origin coming to North America via Iceland and Greenland had little impact on the Native Americans and the Viking settlement at 'Vinland', in the vicinity of Newfoundland and the Gulf of St Lawrence, (Ingstad & Stine Ingstad, 2001) did not survive.

In South America, however, the Incas independently discovered smelting of copper and other metals and making alloys. Gold was used decoratively in large quantities, something that made the culture especially prone to European exploitation, being overthrown and effectively destroyed in a single military campaign by Spanish *Conquistadores*.

During the 18th and 19th Centuries, the mass import of slaves from Africa into, primarily, the Caribbean, the southern states of North America, and Brazil has resulted in very many people of West African origin throughout the Americas, in addition to the large numbers of immigrants from Europe, particularly Britain and Ireland and later from Italy and Eastern Europe. Large parts of Central and South America were populated by Spanish and Portuguese–speaking immigrants from Iberia.

Europe, the Middle East and the Mediterranean Basin.

The fact that Europe came to be the predominant culture world–wide was not always apparent from its early history. For much of the history of modern humans and for Neanderthals, Europe was more an insignificant backwater; an unimportant peninsula on the western end of Eurasia.

It was populated, probably by *H. heidelbergensis*, if indeed they were distinct from *H. erectus*, who were themselves descendants of migrants from Africa who settled in the Middle East or migrated across the Sahara by the operation of the Saharan Pump. There were certainly bipedal hominids in Europe, as shown by the Happisburgh footprints from Norfolk, England, made between 850,000 and 950,000 years ago, although it is by no means certain who they were made by.

The earliest attested presence of several individuals over several generations is in the *Sima de los Huesos* site at Atapuerca, Spain. As we saw in the preceding chapter, these remains show mixed characteristics of *H. antecessor* (? *H. erectus antecessor*), *H. heidelbergensis* (? *H. erectus heidelbergensis*) and *H. neanderthalensis* (dare I suggest that *H. neanderthalensis* might have been *H. erectus neanderthalensis*?). At least one of the female's remains has yielded DNA containing Denisovan genes. Neanderthals were present in Europe for about 250,000 years

H. sapiens came out of Africa about 44,000 years ago and interbred occasionally with Neanderthals. Some of them went the coastal route to South East Asia, Australia and Melanesia; others went up into Central and Northern Asia and still others turned west into Europe. Within a few tens of thousands of years, *H. sapiens* had entirely replaced *H. neanderthalensis* throughout Europe, the Middle East and the Mediterranean basin, the last Neanderthals probably being in Gibraltar in the extreme south of Spain.

There are suggestions that the first *H. sapiens* to come into Europe did not come via the seemingly obvious route via Asia Minor or through the Caucasus and north of the Black Sea or round the Black Sea coast and up the Danube, though some probably did later. The first migrants might have come via coastal spread around the northern Mediterranean coast, into Iberia and round the Western edge of Europe, even to the British Isles and Ireland.

An interesting piece of evidence that the West of Ireland was colonised by people from Iberia in prehistoric times is the presence of a distinct mtDNA lineage of a species of small snail, *Cepaea nemoralis,* in Western Ireland which is found nowhere else other than the Eastern Pyrenees and Southern France. The assumption is that it was taken there either as food itself or accidentally on vegetation. It is absent from the rest of the British Isles, so an over–land route can be excluded (Grindon & Davison, 2013).

The first modern humans to populate Europe were almost certainly dark–skinned hunter–gatherers but about 4,500 years ago a massive influx of nomadic pastoralists, the Yamnaya, probably from the Central Asian steppes north of the Black Sea, brought both the Indo–European language and the genes for pale skin. As I mentioned in the preceding chapter, Eurasian *H. sapiens* may well have acquired the genes for pale skin from Neanderthals, who had evolved it over the preceding 200,000 years as an adaptation to the cloudy, relatively sunless, northern climate (Gibbons, 2015).

Another study in 2012 suggested that these same people may have brought the genes for lactose tolerance into Europe, so enabling early Europeans to make maximum use of cattle (Vuorisalo, et al., 2012).

Agriculture and metallurgical technology are assumed to have spread into Europe and the Mediterranean areas from their discovery in the Middle East, usually placed in the Tigris–Euphrates area in modern Iraq. Wheat was first cultivated in the 'fertile crescent' running from Mesopotamia north of Arabia to Syria.

The Europe–Middle East region benefitted from a number of plants that proved suitable for cultivation in addition to the cereals wheat, barley, oats and rye – the brassicas, alliums, peas and beans, lettuce, root crops such as parsnips and turnips, fruit such as peaches, apricots, plumbs and apples, etc. The now ubiquitous potatoes, tomatoes, peppers (sweet and chile peppers), aubergines and maize were unknown until imported from the Americas in the 16th Century. Eurasia also benefitted from several species of animals that could be domesticated as beasts of burden and draught as well as for food and wool – cattle, horses, donkeys, sheep and goats, pigs, chickens, ducks and later camels in the Middle East and North Africa.

Like rice and maize, cultivated wheat is a hybrid of several varieties and has been highly selected over the centuries so is a good example of co–evolution. One example of probably purely unintentional selection is the tough stalk or rachis attaching the stalk to the ear in cultivated wheat which makes it easier to harvest. Wild wheat has a fragile rachis which means it falls easily from the ear when ripe. Simply by collecting ears of ripe wheat and taking them back to a dwelling for winnowing and grinding, and storage for next years seed, would have automatically selected those with a stronger rachis.

Domestication and selective breeding of animals, including cats and dogs are all examples of co–evolution. Almost all the megafauna of Eurasia was exterminated or brought to the edge of extinction centuries ago, either by climate change or hunting. Those that survived this mini–mass extinction were those that could be domesticated and farmed rather than hunted in the wild.

Early modern humans were firmly established in Western Europe by 20,000 years ago. The Lascaux cave paintings from about then were made by people with a sophisticated appreciation of nature around them. The cave paintings, often deep in caves where they were not on general display, have been interpreted as an attempt to capture the 'spirits' of the animals, which are almost always prey species of big–game hunters.

In places, quite incidentally, they captured the image of a hybrid between the auroch (a large wild bovine) and the Eurasian bison (Soubrier, et al., 2016). This hybrid is believed to be the wild ancestor of domestic cattle. The stable hybrids became a genetically distinct species which alternated with the bison as the dominant species on the Eurasian steppes, depending on changes in climate.

To dispel a common misconception, the Cro–Magnon were not a distinct species of human and have never been claimed to be so. No paleoanthropologist has ever suggested anything other than that Cro–Magnon was an early European *H. sapiens*. The term describes a culture, not a species. The name comes from the *Abri de Cro-Magnon* (French: rock shelter of Cro–Magnon - big cave in Occitan) near Les Eyzies-de-Tayac-Sireuil in southwestern France, where the first specimen was found. Cro–Magnon are associated with the Aurignacian culture and with the cave paintings at Lascaux and elsewhere in France and Spain.

Although the term 'Cro–Magnon' is often, and incorrectly, applied to any early *H. sapiens* the earliest known fully modern *H. sapiens* remains have been found in Romania at *Peștera cu Oase* near the Danube, which may be the route taken by early modern humans into Europe.

Over the centuries, hunter–gatherers were replaced by pastoralists and settled agriculturalists, or they adopted the new ideas themselves. The beginnings of settled farming appear to have been in the Balkans in about 8,500 years ago. This area was also the site of the earliest evidence of copper smelting about 7,000 years ago. Bronze (an alloy of copper and tin and considerably harder than both) was being produced in the area of the Aegean by about 5,200 years ago.

Iron smelting was probably introduced to the area by the Hittites and had spread to Northern Europe by about 2500 years ago. Early Iron Age Europe was almost certainly a settled, agricultural society based on tribal domination of local areas which gradually became more established political entities. Some of these appear to have been highly organised although they have left few direct records or evidence of urban dwellings. Iron Age people of Southern England, for example, were sufficiently well organised to undertake relatively huge projects such as building Stonehenge and Silbury Hill, both of which would have entailed a well–organised, large labour force and an economy able to supply this non–productive workforce with food, clothing and materials. The building of these monuments suggests a powerful religion and a tribal structure possibly based on priest–kings or at least a powerful priest class (Rubicondior, Rosa, 2014).

The Iron Age saw the rise of Greek and then Roman powers in Eastern and Western Europe to rival those of Persia and Egypt in the Middle East and North Africa respectively. An early Greek expansion under Alexander the Great (actually a Macedonia by birth) spread Hellenistic culture throughout Asia Minor, the Levant, Persia and into Southern Central Asia as far as Afghanistan and Bactria, and Northern India. The native Egyptian dynasties were replaced by a Greek dynasty under the Ptolemies, of which Cleopatra was the last.

Greece was later replaced by Roman domination which expended to the Southern shores of the Mediterranean, Iberia, the Hellenised parts of the Middle East and Asia Minor, the Balkans and Central and parts of North West Europe including southern Britain, bringing a sort of peace (*Pax Romanum*) under which trade could flourish. Rome was never able to extend its control to Arabia or Persia and the Hellenised Eastern Empire remained essentially Greek rather than Roman.

Eventually, the Roman Empire collapsed, primarily in the West, under the attack from Germanic tribes. Iberia and Western Europe came under the domination of Germanic tribes, the Franks and Goths. Incursions by Central Asian people, the Huns were replacing Roman control in the Balkans and, under their protection, Slavic people were migrating into the Balkans and down into Greece.

Turkic people, also from the steppes of Central Asia, were migrating *en masse* through the Caucasus and eventually into Asia Minor where they came up against a new power in the Middle East – Islamic Arabs. Arabic armies rapidly spread out from first Mecca and Medina in the Hijaz and then from new centres in Damascus, Baghdad and Jerusalem, into Persia, Egypt and North Africa and eventually into Iberia, even reaching up into southern France as far as Marseille.

With Roman power gone, Europe regressed into smaller, frequently warring, kingdoms and entered the Dark Ages in which it remained until the age of enlightenment, the discovery of science, printing, organised banking and eventually the beginnings of industrialisation.

The age of exploration began in the 15th Century when European culture began to spread round the world to become the dominant culture it is today. Europeans began to colonise large parts of the world, importing a predominantly European culture and people to the Americas, Australia, New Zealand, Canada, and to a lesser extent, Africa. A number of culturally significant diasporas occurred during this period, at different times, notably the Irish, Scottish, Italian, Polish and another Jewish diaspora into primarily North America from Europe.

Meanwhile, founded on Verangian settlements[2] at what were to become Kiev, Smolensk, Rostov, Novgorod, etc., which coalesced into a unified Principality of Russia, the Slavic people of Eastern Europe were beginning to form organised states capable of standing up to the frequent incursions of nomadic tribes from Central Asia, and eventually being able to subdue these remote areas and extend Slavic (Russian) control across Northern Asia, across the great north–flowing Yenisei, Ob and Lena Rivers to the Pacific, coming up against China across the Amur.

End notes:————————————————————

[1] The woomera is a wooden shaft about two to three feet long with a peg at one end onto which the end of a spear can be fitted. When used it extends the power and range of a thrown spear used as a hunting weapon, enabling the hunter to kill at a distance. Although often thought of as more primitive than a bow and arrow, experiments have shown that the extra energy gained from using a woomera can be up to four times that of a compound bow. The woomera appears to have been used from about 5000 years ago.

[2] In the East, settlers and traders from Sweden known as Verangians, the counterpart of the Vikings from Norway, starting in about 800 CE, began travelling down the Northern Russian rivers from the Baltic, establishing settlements and trading posts at a number of centres.

14. The Meaning of Life

So, what does this mean for you and your story?

You are the end-point of your own genes' evolution. You are the descendant of survivors, each of whom bred successfully and never once failed – for 3.5 billion years!

Think about that for a moment. In a world in which, for very many individuals, an early death and failure to breed were by far the most likely outcome, not one single one of your ancestors failed to produce at least one offspring. If they had failed, your gene-line would have ended there and then.

You are the product of billions of passes through the sieve of selection and at every pass your gene-line passed the fitness test. Your genes are good at surviving; and you are unique in the history of the cosmos. The likelihood of you being alive at all is almost vanishingly small and yet here you are. Never before has anyone with your combination of genes, your collection of atoms and your history existed.

And you never will again.

Almost all your genes have spent much longer being something else than they have being human. Your ancestors were there when Europe and Africa split off from the Americas. They were there as small mammal-like reptiles when dinosaurs ruled the earth. They saw pterodactyls flying overhead. They survived the mass-extinction which ended the dinosaurs' reign and they saw the birds and the bats grow wings and take to the air.

Your ancestors swam in the Cambrian seas and crawled out onto the land as early air-gulping fish destined to become four–legged animals with lungs. Your ancestors lived through the Carboniferous era when dense forests of tree ferns grew in steaming jungles where dragonflies with meter-wide wings flew. They saw the trees fall and form the piles of vegetation destined to be coal as the climate changed and the Carboniferous forests collapsed. They saw the first flowering plants as plants and insects formed their mutual-benefit society.

What Makes You So Special?

Your ancestors lived through the first great toxic waste disaster when the cyanobacteria produced oxygen and triggered a mass extinction; and they learned to turn it to their advantage by evolving aerobic respiration.

Your ancestors were bacteria; maybe they were archaea; they may have been the strange Ediacarans which were the earliest known multi-cellular organisms. In almost every one of your cells, in your genes, you carry a record of your evolution, of the entire human evolution story, and of a great deal of the evolution story of every other living thing.

Your journey through space and time has been an adventure of disasters, adaptation, survival and recovery, many, many times you will have been on the brink of extinction - the fate of 99% of all known ancient species - yet your ancestors survived and because they were good at surviving you are here and now.

You will live for a mere flash in the time-scale of the Universe but in the vast darkness of the cosmos there can surely be few flashes as bright as your bright spark of consciousness.

Be proud. Be very proud. But at the same time be humbled by the enormity of the events which produced you and the fragility of it all.

Stars died and because they died, you live. You are made by stars out of stardust and in a very real sense; because you are made of the same stuff the Universe is made of and are a part of it, there is something even more wonderful about you. Through you, though not just through you, and maybe not just here on this small planet, the Universe has gained self–awareness and can begin to understand itself.

Through you it can stand on the surface of this beautiful little jewel in the cosmos, can look up in awe at itself and think "Wow!"

You're special. You are unique and you were nearly 14 billion years in the making.

That is your story. Enjoy it while it lasts.

Please bear in mind also that every other human being; every mammal; every bird, insect, spider, fish, or worm; every plant; indeed, every other living thing, has made the same journey you have made. Each is unique and the descendant of survivors. Each has an unbroken gene-line going back to the first replicator. They are your relatives. Like you they are part of the whole web of interdependent things we call life on earth.

To end their life will end their gene-line for the first and only time in the history of the Universe. Something which has taken nearly 14 billion years to produce, and 3.5 billion years to perfect, will have been extinguished forever.

Each of them is worthy of respect and each of them deserves the one opportunity to experience life that chance has given them.

Life is too rare, precious and wonderful a thing to take lightly.

And the meaning of life?

The meaning of life is whatever meaning you want to give your life. It is your life. Do with it whatever you want with it but try not to waste it.

What Makes You So Special?

Bibliography

Adami, C., & Hintze, A. (2013, Aug 1). Evolutionary instability of zero-determinant strategies demonstrates that winning is not everything. *Nature Communications, 4*, Article No. 2193. doi:10.1038/ncomms3193

Anderson, J., Smithson, T., Mansky, C., Mayer, T., & Clark, J. (2015, Apr 27). A Diverse Tetrapod Fauna at the Base of 'Romer's Gap'. *PLOS One, 10*(4). doi:10.1371/journal.pone.0125446

Ashton, N., Lewis, S. G., De Groote, I., Duffy, S. M., Bates, M., Bates, R., . . . Stringer, C. (2014, Feb 7). Hominin Footprints from Early Pleistocene Deposits at Happisburgh, UK. *PLOS ONE, 9*(2), e88329. doi:10.1371/journal.pone.0088329

Balaresque, P., Poulet, N., Cussat-Blan, S., Gerard, P., Quintana-Murc, L., Heyer, E., & Jobling, M. A. (2015, Jan 14). Y-chromosome descent clusters and male differential reproductive success: young lineage expansions dominate Asian pastoral nomadic populations. *European Journal of Human Genetics, 23*, 1413–1422. doi:10.1038/ejhg.2014.285

Beach, D. (1998, Feb). Cognitive Archaeology and Imaginary History at Great Zimbabwe. *Current Anthropology, 39*(1), 47-72 . doi:10.1086/204698

Bienvenu, T., Falk, D., Semendeferi, K., Guy, F., Zollikofer, C., Ponce De León, M., . . . Brunet, M. (2013). The endocast of Sahelanthropus tchadensis, the earliest known hominid (7 Ma, Chad). *The 82nd Annual Meeting of the American Association of Physical Anthropologists (2013)*. Retrieved May 23, 2017, from http://meeting.physanth.org/program/2013/session16/bienvenu-2013-the-endocast-of-sahelanthropus-tchadensis-the-earliest-known-hominid-7-ma-chad.html

Böhme, M., Spassov, N., Ebner, M., Geraads, D., Hristova, L., Kirscher, U., . . . Winklhofer, M. (2017, May 22). Messinian age and savannah environment of the possible hominin Graecopithecus from Europe. *PLOS ONE, 12*(5), e0177347. . doi:10.1371/journal.pone.0177347

Bonner, J. T. (1998). The Origins of Multicellularity. *Integrative Biology, 1*(1), 27–36. doi:10.1002/(SICI)1520-6602(1998)1:1<27::AID-INBI4>3.0.CO;2-6

Callaway, E. (2013, Dec 5). Mystery humans spiced up ancients' sex lives. *Nature, 504*, pp. 16-17. doi:10.1038/nature.2013.14196

Carey, N. (2012). *The Epigeenetics Revolution: How Modern Biology is Rewriting our Understanding of Genetics, Disease and Inheritance* (Reprint ed.). London: Icon Books. Retrieved May 15, 2017, from https://www.amazon.co.uk/Epigenetics-Revolution-Rewriting-Understanding-Inheritance/dp/1848313470

Chauhan, P. R. (2003). *An Overview of the Siwalik Acheulian & Reconsidering Its Chronological Relationship with the Soanian – A Theoretical Perspective*. Retrieved May 26, 2017, from assemblage: https://archaeologydataservice.ac.uk/archives/view/assemblage/html/7/chauhan.html

Civáň, P., Craig, H., Cox, C. J., & Brown, T. A. (2015, Nov 2). Three geographically separate domestications of Asian rice. *Nature Plants, 1*, Article #15164. doi:10.1038/nplants.2015.164

Conselice, C. J., Wilkinson, A., Duncan, K., & Mortlock, A. (2016, Oct 14). The Evolution Of Galaxy Number Density At Z < 8 And Its Implications. *The Astrophysical Journal, 830*(2). doi:10.3847/0004-637X/830/2/83

Court Transcript. (2005, Oct 19). *Day 12 AM (PDF) Page 22 Ln 20 - Page 23 Ln 5*. Retrieved May 10, 2017, from National Council for Science Education - Kitzmiller Trial Transcripts: https://ncse.com/files/pub/legal/kitzmiller/trial_transcripts/2005_1019_day12_am.pdf

Dawkins, R. (2006). *Unweaving the Rainbow: Science, Delusion and the Appatite for Wonder*. London: Penguin Books.

Dawkins, R., & Wong, Y. (2006). *The Ancstor's Tale: A Pilgrimage to the Dawn of Life*. London: Weidenfeld & Nicholson.

de Manuel, M., & et al. (2016, Oct 28). Chimpanzee genomic diversity reveals ancient admixture with bonobos. *Science, 354*(6311), 477-481. doi:10.1126/science.aag2602

Discovery Institute. (Undated). *The Wedge Document*. Retrieved May 10, 2017, from National Council for Science Education - The Wedge: https://ncse.com/files/pub/creationism/The_Wedge_Strategy.pdf

Evans, R. (2015). *The Cosmic Microwave Background: How It Changed Our Understanding of the Universe*. New York: Springer International Publishing.

Finlayson, C., Giles Pacheco, F., Rodriguez-Vidal, J., Fa, D. A., Maria Gutierrez Lopez, J., Santiago Perez, A., . . . Fernandez Jalvo, Y. (2006, Oct 19). Late survival of Neanderthals at the southernmost extreme of Europe. *Nature, 443*, 850-853. doi:10.1038/nature05195

Fu, Q., Li, H., Moorjani, P., Jay, F., Slepchenko, S., Bondarev, A., . . . Do. (2014, October 23). Genome sequence of a 45,000-year-old modern human from western Siberia. *Nature, 514*(7523), 445–449. doi:10.1038/nature13810

Gibbons, A. (2015, Apr 2). How Europeans evolved white skin. *Science*. doi:10.1126/science.aab2435

Gould, S. J. (1990). *Wonderful Life: Burgess Shale and the Nature of History*. London: Random House.

Grindon, A. J., & Davison, A. (2013, Jun 19). Irish *Cepaea nemoralis* Land Snails Have a Cryptic Franco-Iberian Origin That Is Most Easily Explained by the Movements of Mesolithic Humans. *PLOS ONE, 8*(6), e65792. doi:10.1371/journal.pone.0065792

Grosberg, R. K., & Strathman, R. R. (2007, December). The evolution of multicellularity: A minor major transition? *Annual Review of Ecology, Evolution, and Systematics, 38*, 621–654. doi:10.1146/annurev.ecolsys.36.102403.114735

Haig, D. (2012, August 7). Retroviruses and the Placenta. *Current Biology, 22*(15), R609–R613. doi:10.1016/j.cub.2012.06.002

Hawking, S. (1994). *Black Holes And Baby Universes And Other Essays.* London: Bantam Books.

Huerta-Sanchez, E., Jin, X., Asan, Bianba, Z., Peter, B. M., Vinckenbosch, N., . . . Wang, W. e. (2014, Aug 14). Altitude adaptation in Tibetans caused by introgression of Denisovan-like DNA. *Nature, 512*, 194–197. doi:10.1038/nature13408

Ingstad, H., & Stine Ingstad, A. (2001). *The Viking Discovery of America: The Excavation of a Norse Settlement in L'Anse Aux Meadows, Newfoundland.* Facts On File Inc.

Jinzhuang, X., Zhenzhen, Deng, Pu, Huang, Kangjun, Huang, Michael J., Benton, Ying, Cui, . . . Shougang, Hao. (2016, August 8). Belowground rhizomes in paleosols: The hidden half of an Early Devonian vascular plant. *PNAS, 113*(34), 9451-9456. doi:10.1073/pnas.1605051113

Krauss, L. M. (2012). *A Universe From Nothing: Why There Is Something Rather Than Nothing.* New York: Free Press. Retrieved May 15, 2017, from https://www.amazon.com/Universe-Nothing-There-Something-Rather/dp/1451624468/

Lahaye, C., Hernandez, M., Boëda, E., Felice, G. D., Guidon, N., Hoeltz, S., . . . Viana, S. (2013, June). Human occupation in South America by 20,000 BC: the Toca da Tira Peia site, Piauí, Brazil. *Journal of Archaeological Science, 40*(6), 2840–2847. doi:10.1016/j.jas.2013.02.019

Lane, N., & Le Page, M. (2009, Oct 14). *How life evolved: 10 steps to the first cells*. Retrieved from New Scientist: https://www.newscientist.com/article/dn17987-how-life-evolved-10-steps-to-the-first-cells/

Lazaridis, I., Patterson, N., Mittnik, A., Renaud, G., Mallick, S., Kirsanow, K., . . . et al. (2014, Sep 18). Ancient human genomes suggest three ancestral populations for present-day Europeans. *Nature, 513*, 409–413. doi:10.1038/nature13673

Lee, M. S., Soubrier, J., & Edgecombe, G. D. (2013, Sep 12). Rates of Phenotypic and Genomic Evolution during the Cambrian Explosion. *CurrentBiology, 23*(19), 1889 - 1895. doi:10.1016/j.cub.2013.07.055

Margulis, L. (1998). *Symbiotic Planet: A New Look At Evolution* (Revised ed.). Massachusetts: Basic Books.

Martin, T., Marugán-Lobón, J., Vullo, R., Martín-Abad, H., Zhe-Xi , L., & Buscalioni, A. D. (2015, Oct 15). A Cretaceous eutriconodont and integument evolution in early mammals. *Nature, 526*, 380–384. doi:10.1038/nature14905

Masao, F. T., Ichumbaki, E. B., Cherin, M., Barili, A., Boschian, G., Iurino, D. A., . . . Manzi, G. (2016, Dec 14). New footprints from Laetoli (Tanzania) provide evidence for marked body size variation in early hominins. *eLife, 5*. doi:10.7554/eLife.19568.001

Mendez, F. L., Poznik, G. D., Castellano, S., & Bustamante, C. D. (2016, Apr 7). The Divergence of Neandertal and Modern Human Y Chromosomes. *The American Journal of Human Genetics, 98*(4), 728 - 734. doi:10.1016/j.ajhg.2016.02.023

Nancy, P., Tagliani, E., Chin-Siean, T., Asp, P., Levy, D. E., & Erlebacher, A. (2012, Jun 8). Chemokine Gene Silencing in Decidual Stromal Cells Limits T Cell Access to the Maternal-Fetal Interface. *Science, 336*(6086), 1317-1321. doi:10.1126/science.1220030

Niklas, K. J. (2013, Nov). The evolutionary-developmental origins of multicellularity. *American Journal of Botany, 101*(1), 6-25. doi:10.3732/ajb.1300314

Oppenheimer, S. (2003). *Out of Eden: The Peopling of the World.* London: Constable & Robinson. Retrieved May 27, 2017, from https://www.amazon.com/Out-Eden-Peopling-Stephen-Oppenheimer/dp/1841196975

Oskarsson, M. C., Klütsch, C. F., Boonyaprakob, U., Wilton, A., Tanabe, Y., & Savolainen, P. (2012, Mar 7). Mitochondrial DNA data indicate an introduction through Mainland Southeast Asia for Australian dingoes and Polynesian domestic dogs. *Proc Biol Sci, 279*(1730), 967–974. doi:10.1098/rspb.2011.1395

Parfrey, L. W., & Lahr, D. J. (2013, Apr). Multicellularity arose several times in the evolution of eukaryotes (Response to DOI 10.1002/bies.201100187). *BioEssays, 35*(4), 339–347. doi:10.1002/bies.201200143

Patterson, N., Richter, D. J., Gnerre, S., Lander, E. S., & Reich, D. (2006, Jun 29). Genetic evidence for complex speciation of humans and chimpanzees. *Nature, 441*, 1103-1108. doi:10.1038/nature04789

Peng, Y., Shi, H., Qi, X.-b., Xiao, C.-j., Zhong, H., Ma, R.-l. Z., & Su, B. (2010, Jan 20). The ADH1B Arg47His polymorphism in East Asian populations and expansion of rice domestication in history. *BMC Evolutionary Biology, 10*(1), 15. doi:10.1186/1471-2148-10-15

Polkinhorne, J. (2002). *Quantum Theory: A Very Short Introduction.* Oxford: Oxford University Press.

Pringle, *H.* (2009, Jan 9). Archaeologists are battling over when--and how--ancient African cultures entered the Iron Age. *Science, 323*(5911), pp. 200-202. doi:10.1126/science.323.5911.200

Rasmussen, M., Anzick, S. L., Waters, M. R., Skoglund, P., DeGiorgio, M., Stafford Jr, T. W., & et al. (2014, Feb 13). The genome of a Late

Pleistocene human from a Clovis burial site in western Montana. *Nature, 506*, 225–229. doi:10.1038/nature13025

Rubicondior, R. (2013, Nov 10). *Lessons From Nature.* Retrieved May 19, 2017, from Rosa Rubicondior: http://rosarubicondior.blogspot.co.uk/2013/11/lessons-from-nature.html

Rubicondior, Rosa. (2010, Sep 27). *What Makes You So Special?* Retrieved May 10, 2017, from Rosa Rubicondior: http://rosarubicondior.blogspot.co.uk/2010/09/what-makes-you-so-special.html

Rubicondior, Rosa. (2011, Jul 31). *Why Species?* Retrieved May 12, 2017, from Rosa Rubicondior: http://rosarubicondior.blogspot.co.uk/2011/07/why-species.html

Rubicondior, Rosa. (2012, Jul 04). *Evolution - Making a Monkey.* Retrieved May 12, 2017, from Rosa Rubicondior: http://rosarubicondior.blogspot.co.uk/2012/07/evolution-making-monkey-of-creationists.html

Rubicondior, Rosa. (2012, May 01). *The Good Shepherd's Purse is Bad News For Creationists.* Retrieved May 09, 2017, from Rosa Rubicondior: http://rosarubicondior.blogspot.co.uk/2012/05/good-shepherds-purse-is-bad-news-for.html

Rubicondior, Rosa. (2013, May 02). *Looking at Life.* Retrieved May 09, 2017, from Rosa Rubicondior: http://rosarubicondior.blogspot.co.uk/2011/10/what-is-life.html

Rubicondior, Rosa. (2014, Oct 14). *Old Dead Gods - Lessons from Silbury Hill.* Retrieved May 31, 2017, from Rosa Rubicondior: http://rosarubicondior.blogspot.co.uk/2014/10/old-dead-gods-lessons-from-silbury-hill.html

Soares, P., Rito, T., Trejaut, J., Mormina, M., Hill, C., Tinkler-Hundal, E., . . . Richards, M. B. (2011, Feb 11). Ancient Voyaging and Polynesian

Origins. *The American Journal of Human Genetics, 88*(2), 239 - 247.
doi:10.1016/j.ajhg.2011.01.009

Soubrier, J., Gower, G., Chen, K., Richards, S. M., Llamas, B., Mitchell, K. J.,
. . . Fordham, D. A. (2016, Oct 18). Early cave art and ancient DNA
record the origin of European bison. *Nature Communications, 7*,
Article #13158. doi:10.1038/ncomms13158

Stephens, N. J., Seiffert, E. R., O'Connor, P. M., Roberts, E. M., Schmitz, M.
D., Krause, C., . . . Temu, J. (2013, May 30). Palaeontological
evidence for an Oligocene divergence between Old World monkeys
and apes. *Nature, 497*, 611–614. doi:10.1038/nature12161

The Burgess Shale Site 510 Million Years Ago. (n.d.). Retrieved May 14, 2017,
from Smithsonian:
http://paleobiology.si.edu/burgess/cambrianWorld.html

Vuorisalo, T., Arjamaa, O., Vasemägi, A., Taavitsainen, J.-P., Tourunen, A., &
Saloniemi, I. (2012). High lactose tolerance in North Europeans: a
result of migration, not in situ milk consumption. *Perspectives in
Biology and Medicine, 55*(2), 163-174. doi:10.1353/pbm.2012.0016

Wikipedia. (n.d.). *Multicellular organism.* Retrieved May 13, 2017, from
Wikipedia: https://en.wikipedia.org/wiki/Multicellular_organism

Index

Books by Rosa Rubicondior

The Light of Reason Series:

The Light of Reason: And Other Atheist Writings.
Irreverent essays, thought-provoking articles and humorous items on atheism, religion, science, evolution, creationism and related issues.

(Hardcover|) ISBN-13: 979-8512173916	£13.50 (US $18.50)
(Paperback) ISBN-10: 1516906888, ISBN-13: 978-1516906888	£9.95 (US $14.95)
(Kindle) ASIN: B014N0IPVI	£3.95 (US $5.99)

The Light of Reason: Volume II – Atheism, Science and Evolution.
Thought-provoking essays on the conflict between fundamentalist religion and science, and exposing the anti-science, extremist political agenda of the modern creationist industry.

(Hardcover) ISBN-13: 979-8512191040	£13.50 (US $18.50)
(Paperback) ISBN-10: 1517105188, ISBN-13: 978-1517105181	£9.95 (US $14.95)
(Kindle) ASIN: B014N0IR16	£3.99 (US $5.99)

The Light of Reason: Volume III – Apologetics, Fallacies, and Other Frauds.
Thought-provoking essays and articles on religion and atheism, dealing with religious apologetics, fallacies, miracles and other frauds

(Hardcover) SBN-13: 979-8512526002	£12.00 (US $17.00)
(Paperback) ISBN-10: 151710761X, ISBN-13: 978-1517107611	£6.95 (US $9.95)
(Kindle) ASIN: B014N0IRE8	£2.99 (US $3.99)

The Light of Reason: Volume IV - The Silly Bible.
Exposing the absurdities, contradictions and historical inaccuracies in the Bible and advancing the case for atheism and against religion. This volume, the fourth in the Light of Reason series, deals with contradictions and absurdities in the Bible.

(Hardcover) ISBN-13: 979-8512539392	£13.50 (US $18.50)
(Paperback) ISBN-10: 1517108209, ISBN-13: 978-1517108205	£8.95 (US $13.95)
(Kindle) ASIN: B014N0IR8E	£3.99 (US $4.99)

The Light of Reason: And Other Atheist Writing. (all 4 volumes in one e-book)
Based on the Rosa Rubicondior science and Atheism blog, this is a collection of Atheist and science articles, some short, others lengthier, exploring the interface between religion and science and which have been published over some four years.

(Kindle only) ASIN: B013DYOK32	£6.34 (US $9.95)

Other books on science, Atheism and theology

An Unprejudiced Mind: Atheism, Science & Reason.
Essays on science and theology from a scientific atheist perspective, exploring particularly evolution versus creationism.

(Hardcover) ISBN-13: 979-8512554685	£13.10 (US $18.50)
(Paperback) ISBN-10: 1522925805, ISBN-13: 978-1522925804	£9.95 (US $14.95)
(Kindle) ASIN: B019UGXPM4	£3.99 (US $5.95)

Ten Reasons To Lose Faith: And Why You Are Better Off Without It.
Why faith is not only a fallacy and useless as a route to the truth but is actually harmful to society and to the individual. It systematically dismantles the standard religious apologetics and shows them to be bogus and deliberately constructed to mislead.

(Hardcover) ISBN-13: 979-8509108433	£16.00 (US $22.00)
(Paperback). ISBN-13:978-1530431953, ISBN–10: 1530431956	£10.75 (US $14.75)
(Kindle) ASIN: B01DGVO3JS	£6.95 (US $8.95)

What Makes You So Special? : From the Big Bang to You.
How did you come to be here, now? This book takes you from the Big Bang to the evolution of modern humans and the history of human cultures

(Hardcover) ISBN-13: 979-8509108433	£13.50 (US $18.00)
(Paperback) ISBN-13: 978-1546788294, ISBN-10: 1546788298	£8.95 (US $11.50)
(Kindle).ASIN: B071FTKXLZ	$6.20 (US $8.25)

The Internet Handbooks series

The Internet Creationists' Handbook: A Joke for the Rest of Us.
A humorous look at creationist apologetics on the Internet, showing the fallacies and dishonest tactics creationists are using to try to recruit scientifically illiterate people into their political cult.

(Paperback),ISBN-13: 978-1721605149, ISBN-10: 1721605149£5.25 (US $7.50)	
(Kindle) ASIN: B07DZF75KD	£3.75 (US $5.00)

The Christian Apologists' Handbook: A Joke for the Rest of Us.
A humorous look at Christian apologetics on the Internet, showing the fallacies and dishonest tactics Christian fundamentalists are using to try to recruit scientifically and theologically illiterate people to their cults, often with political motives.

(Paperback) ISBN-13: 978-1721724727, ISBN–10: 1721724729	£5.25 (US $7.50)
(Kindle) ASIN: B07DYDVMW4	£3.75 (US $5.00)

Books by Rosa Rubicondior

The Muslim Apologists' Handbook: A Joke for the Rest of Us.
  A humorous look at Muslim apologetics on the Internet, showing the fallacies and dishonest tactics
  Muslim fundamentalists are using to try to recruit scientifically and theologically illiterate people to their
  cuts, often with political motives.

(Paperback) ISBN-13: 978-1721756896, ISBN-10: 1721756892	£5.25 (US $7.50)
(Kindle) ASIN: B07DZF75KD	$3.75 (US $5.00)

The Unintelligent Design Series

The Unintelligent Designer: Refuting the Intelligent Design Hoax
  Showing why the superficial appearance of design in living things cannot be attributed to anything
  like an intelligent designer, as a counter to the politically-motivated Intelligent Design movement.

(Hardcover) ISBN-13: 979-8513528463	£13.50 (US $18.50)
(Paperback) ISBN-10: 1723144215, ISBN-13: 978-1723144219	£9.00 (US $12.50)
(Kindle) ASIN B07G121BMK	£5.00 (US $7.00)

The Malevolent Designer: Why Nature's God is not Good
  Showing why, if we accept for the sake of argument the Creationist insistence on Intelligent Design
  as the best explanation for biodiversity on Earth, the creator god they purport to worship could not
  be regarded as anything other than a malevolent evil, assiduously designing suffering into its
  creation as though it hates it and wants it to suffer in unimaginably horrible ways.

Illustrated by Catherine Hounslow-Webber

(Hardcover) ISBN-13: 979-8511295442	£15.00 (US $18.00)
(Paperback) SBN-13; 979-8670361729	£9.10 (US $12.50)
(Kindle) ASIN: B08L9S8F5F	£5.50 (US $7.60)

Publish under the name Bill Hounslow – Oxfordshire Childhood series.

In The Blink of an Eye: Growing Up in Rural Oxfordshire
  A frank recollections of life as feral children in the small North Oxfordshire hamlet of Fawler during
  the 1950s and 60s, on the brink of major change as we approached the television age and the final
  stages in the domestication of children was about to begin.

Additional material by Patricia Broome

(Hardcover) ISBN-13: 979-8511967400	£13.00 (US $18.50)
(Paperback) ISBN-10: 1545350787, ISBN-13: 978-1545350782	£6.50 (US $11.49)
(Kindle) ASIN: B06ZY8JZ92	$6.50 (US $8.95)

A Goose for Christmas: Stories from an Oxfordshire Childhood

Slightly imaginative stories, based on real events and people, of childhood adventures in the North Oxfordshire hamlet of Fawler in the 1950s during the post-war austerity, before television, when the children had only what they could get from the woods and fields around them.

Illustrated by Catherine Webber-Hounslow

(Hardcover) ISBN-13: 979-8511907482   £12.75 (US $18.00)
(Paperback) ISBN-13: 978-1981708925, ISBN-10: 1981708928   £8.50 (US $11.50)
(Kindle) ASIN: B07GFJ85P8